AF334998

# QUANTUM THEORY AND GRAVITATION

**Academic Press Rapid Manuscript Reproduction**

Proceedings of a symposium held at Loyola University,
New Orleans, May 23–26, 1979.

# QUANTUM THEORY AND GRAVITATION

edited by

## A. R. MARLOW

*Department of Physics*
*Loyola University*
*New Orleans, Louisiana*

ACADEMIC PRESS    1980

A Subsidiary of Harcourt Brace Jovanovich, Publishers

New York   London   Toronto   Sydney   San Francisco

COPYRIGHT © 1980, BY ACADEMIC PRESS, INC.
ALL RIGHTS RESERVED.
NO PART OF THIS PUBLICATION MAY BE REPRODUCED OR
TRANSMITTED IN ANY FORM OR BY ANY MEANS, ELECTRONIC
OR MECHANICAL, INCLUDING PHOTOCOPY, RECORDING, OR ANY
INFORMATION STORAGE AND RETRIEVAL SYSTEM, WITHOUT
PERMISSION IN WRITING FROM THE PUBLISHER.

ACADEMIC PRESS, INC.
111 Fifth Avenue, New York, New York 10003

*United Kingdom Edition published by*
ACADEMIC PRESS, INC. (LONDON) LTD.
24/28 Oval Road, London NW1    7DX

**Library of Congress Cataloging in Publication Data**

Main entry under title:

Quantum theory and gravitation.

    Proceedings of a symposium held at Loyola
University, New Orleans, May 23–26, 1979.
    1.  Quantum theory—Congresses.    2.  Gravitation
—Congresses.    I.   Marlow, A. R.
QC173.96.Q82          530.1′2          79-27783
ISBN 0-12-473260-7

PRINTED IN THE UNITED STATES OF AMERICA

80 81 82 83     9 8 7 6 5 4 3 2 1

# CONTENTS

# CONTRIBUTORS

Numbers in parentheses refer to the pages on which authors' contributions begin.

Jeeva S. Anandan (157), Center for Theoretical Physics, Department of Physics and Astronomy, University of Maryland, College Park, Maryland 20740

Carroll F. Blakemore (233), Department of Mathematics, University of New Orleans, New Orleans, Louisiana 70122

Carl H. Brans (27), Department of Physics, Loyola University, New Orleans, Louisiana 70118

Lutz Castell (147), Max-Planck-Institut zur Erforschung der Lebensbedingungen, der wissenschaftlich-technischen Welt, Riemerschmidstrasse 7, Postfach 1529, D-8130 Starnberg, Germany

George F. Chapline (177), Theoretical Physics Division, Lawrence Livermore Laboratory, University of California, Livermore, California 94550

Maurice J. Dupré (199), Department of Mathematics, Tulane University, New Orleans, Louisiana 70118

David Finkelstein (79), School of Physics, Georgia Institute of Technology, Atlanta, Georgia 30332

S. A. Fulling (187), Department of Mathematics, Texas A&M University, College Station, Texas 77843

Jerome A. Goldstein (207), Department of Mathematics, Tulane University, New Orleans, Louisiana 70118

Robert Hermann (95), Division of Applied Sciences, Harvard University, Cambridge, Massachusetts 02138

Arthur Komar (127), Department of Physics, Yeshiva University, New York, New York 10033

A. R. Marlow (35, 71), Department of Physics, Loyola University, New Orleans, Louisiana 70118

J. G. Miller (221), Department of Mathematics, Texas A&M University, College Station, Texas 77843

Phillip E. Parker (137), Department of Mathematics, Syracuse University, Syracuse, New York 13210

John Archibald Wheeler (1), Center for Theoretical Physics, The University of Texas at Austin, Austin, Texas 78712

W. K. Wootters (13), Center for Statistical Mechanics and Center for Theoretical Physics, The University of Texas at Austin, Austin, Texas 78712

# PREFACE

This volume contains the results of the second in a series of meetings at Loyola in New Orleans between physicists and mathematicians concerned with the fundamental questions of modern theoretical physics. The first conference in June 1977 centered around foundational problems in quantum theory. Emboldened by the success of that meeting, the conference organizers (C. H. Brans and A. R. Marlow) decided to cast their net in a wider arc and invite a distinguished group of participants to focus their collective talents in May 1979 on the unification of the two theories that define twentieth century physics: quantum theory and general relativity.

The diversity of the participants is reflected in the diversity of the final results presented in this volume: overviews designed to locate and clearly define the problem areas, specific solutions to problems in quantum theory or relativity, frontal attacks on the central question itself. One of the premises from which a conference of this type arises is the belief that mathematicians and physicists talking and working together can accomplish more than either group separately. We believe this premise has been sustained. In particular, the formative influence on the shape of the final results contributed by a distinguished group of stimulators, facilitators, constructive critics, and just plain good listeners (Andrew M. Gleason, R. J. Greechie, Paul R. Halmos, Cecile DeWitt-Morette, Bryce S. DeWitt) should not go unrecognized. Throughout the conference the participation of John A. Wheeler, the one figure in modern physics who has perhaps contributed most definitively to our present understanding of both quantum theory and gravitation, acted as a yeast to make the whole thing rise.

Albert Einstein founded relativity theory and contributed mightily to quantum theory. The success of the present efforts to unify the two theories in the centennial year of his birth must be left to the judgment of the reader. As a possible guide through this volume, one might read first Wheeler's survey of pregeometry as foundation, then Brans' study of the problem areas, then Marlow's construction of a relativistic quantum model from quantum logic as pregeometry, and finally use Blakemore's conference survey (last paper) as a key to the full variety of the contributions.

Needless to say, the contributions of so many persons are needed to make a conference and publication of this type come about that no listing can be adequate. However, we feel that we must attempt to repay in some small way our debt of gratitide to the following individuals and groups: Research Corporation (matching funds grant), James C. Carter, S.J., President, Robert A. Preston, Academic Vice President, John F. Christman, Director of Research, Loyola University (Loyola research funding); Henry Montecino, S.J., and the members of the Loyola Jesuit Community (financial and moral support); Creston A. King, Jr., Francis A. Benedetto, S.J., David G. Keiffer and Henry A. Garon, Loyola Department of Physics (financial and moral support). Among those whose services were essential to the success of the conference and this volume are: Germaine L. Murray, Keith D. Bonin, Marian V. Bonin, Louis M. Barbier, Carlos Barrera, Jaime Ayarza, John R. Blum, and Ray Erato. Lastly, the assistance of the editor and staff of Academic Press is deeply appreciated in bringing this volume to its final form.

# QUANTUM THEORY AND GRAVITATION

# PREGEOMETRY:  MOTIVATIONS AND PROSPECTS

John Archibald Wheeler[1]

Center for Theoretical Physics
University of Texas
Austin, Texas

Spacetime is normally treated as a continuum.  Likewise for many purposes, an elastic medium is usefully regarded as a continuum or a piece of cloth as continuum.  However, an elastic substance reveals at a crack that the concept of "ideal elastic medium" is a fiction.  Cloth shows at a sel-vage that it is not a continuous medium but woven out of thread.  Spacetime--with or without "gauge" or "phase" or "internal spin" degrees of freedom--often considered to be the ultimate continuum of physics, evidences nowhere more clearly than at big bang and at collapse that it cannot be a continuum.  Obliterated in those events is not only matter, but the space and time that envelop that matter.  If the elastic medium is built out of electrons and nuclei and nothing more, if cloth is built out of thread and nothing more, we are led to ask out of what "pregeometry" the geometry of space and spacetime are built.  What are the motives for asking about "pregeometry;" and what the clues?

Investigators of an earlier age often asked, "Why does space have dimension three?"  Many interesting proposals were

[1]Publication assisted by this center and by NSF grant PHY78-26592.



Copyright © 1980 by Academic Press, Inc.
All rights of reproduction in any form reserved.
ISBN 0-12-473260-7

put forward in response to this question (Kant, 1868; Poincaré, 1913; Ehrenfest, 1917; Reichenbach, 1928; Weyl, 1963).  In the end, however, we can believe it is the wrong question for understanding space.  Might we not more appropriately ask, "How does the world manage to give the impression that it has dimension three?"

A "homogeneous isotropic elastic substance" manages to give the impression that it has two and only two elastic constants.  Nevertheless, a closer look shows that the very concept of elasticity is only an approximation.  There is no such thing as "elasticity" in the space between the electron and the nucleus.  Moreover, a hundred years of the study of elasticity would never have revealed atoms, molecules and the complicated dependence on distance of the forces that hold them together.  Nor would a hundred years of the study of atomic and molecular forces have revealed that they go back for their origin to electrons, nuclei and Schrödinger's equation and nothing more.  The direction of understanding went, not from the large to the small, but from the small to the large.  If elasticity was  the last place to look for a clue to Schrödinger's equation, geometry would seem the last place to look for a clue to pregeometry.

It is possible to dispose of two unproductive ideas of pregeometry--a lattice and Borel set--and yet confess the total absence today of any productive idea of pregeometry.

The word "dozens" is too small to describe the number of papers which treat space as a lattice.  They include (1) Heisenberg's (1930) consideration and rejection of a lattice geometry as a way to deal with the self-energy difficulty of electron theory, (2) Snyder's (1947a, 1947b) proposal and development of relativistic commutation relations between space and time coordinates, (3) Regge's (1961) skeletonization of Riemannian geometry as means both to bring out the geometric content of Einstein's standard geometrodynamics and as algorithm for calculating the evolution of geometry with time, and (4) treating field theory in general, whether electrodynamics or Yang-Mills field or other fields, as taking place on a lattice of points rather than on a continuum, as a means to make tractable the mathematical analysis of some of those field theories (Maxwell, 1877; Kogut, 1978). All of these investigations treat spacetime as a pre-existing continuum; they do not look at it as an approximation to an underlying structure, a pregeometry, a substrate of quite a different kind.

If it does not lead us to pregeometry to accept already in advance a space, or a spacetime continuum of a definite dimensionality, and skeletonize it, what about the opposite approach? Can one take a set of points, a so-called Borel set, that in the beginning has no dimensionality, assemble these points into the most diverse configurations and look for a consideration of probability, by way of Feynman's (Feynman, 1942; Feynman and Hibbs, 1965) sum over histories or otherwise, that will give preference to three dimensions as compared to other dimensions (Wheeler, 1964)? Here also too much geometric structure is presupposed to lead to a believable theory of geometric structure. No one has ever come forward with a way to ascribe a weight or a probability amplitude to the configuration of a Borel set of points that did not rest on some idea of distance between point and point.  But to admit distance at all is to give up on the search for pregeometry.

At least one try has been made at a concept of pregeometry that breaks loose at the start from all mention of geometry and distance, "Pregeometry as the calculus of propositions" (Wheeler, 1971; Misner et al., 1973):  "...make a statistical analysis of the calculus of propositions in the limit where the number of propositions is great and most of them are long. Ask if parameters force themselves on one's attention in this analysis (1) analogous in some small measure to the temperature and entropy of statistical mechanics but (2) so much more numerous, and everyday dynamic in character, that they reproduce the continuum of everyday physics."  However, a later analysis (Patton, 1975) found nothing in mathematical logic supportive of this proposal.  On the contrary, what is most attractive in mathematical logic--the theorem of Gödel (1931, 1934) and related theorems (Cohen, 1966) is the least attractive.  The modern revolution in mathematical logic points, not toward some chosen branch of mathematical logic as the natural foundation for pregeometry and physics, but away.  The Gödel theorem of undecidability shows up in number theory, in the theory of transfinite members, in set theory and in any formal axiomatic system of more than minimal complexity.

In the end we are led back from mathematics to physics in the search for a clue to pregeometry.  The only thing that could be worse than not finding pregeometry automatically contained in mathematics would be finding it automatically

contained in mathematics.  How could one believe any account
of the foundation for the central structure of physics,
spacetime, which proceeded without reference to the quantum,
the overarching principle of all physics?

The central lesson of the quantum has been stated in the
words, "No elementary phenomenon is a phenomenon until it is
an observed (registered) phenomenon" (Wheeler, 1979).  Nowhere
does this feature of nature show more conspicuously than in
so-called "delayed-choice" experiments (Wheeler, 1978).  Nowhere is this "question and answer" way of converting con-
ceivabilities into actualities illustrated in a more homely
context than in the "surprise version" of the game of twenty
questions (Wheeler, 1978).  There one sees the one who thought
he was an observer pure and simple willy-nilly converted into
a participator.  Both in the game and in the elementary
quantum phenomenon the observer-participator converts con-
ceivability into actuality.  If at this elementary level we
already have a mechanism for building part of what we call
reality, why should we look further for a mechanism to build
all of what we call reality--including spacetime itself?
Does not the famous "razor" of Duns Scotus and William of
Occam, "Essentia non sunt multiplicanda praeter necessitatem"
instruct us not to look for two methods of constructing
reality when we already have one?

That is the task; what is the vision (Wheeler, 1979)?
(1) Law without law with no before before the big bang and
no after after collapse.  The universe and the laws that
guide it could not have existed from everlasting to everlasting.  Law must have come into being (Peirce, 1940).

Moreover, there could have been no message engraved in advance on a tablet of stone to tell them how to come into being. They had to come into being in a higgledy-piggledy way, as the order of genera and species came into being by the blind accidents of billions upon billions of mutations, and as the second law of thermodynamics with all its dependability and precision comes into being out of the blind accidents of motion of molecules who would have laughed at the second law if they had ever heard of it. (2) "Individual events. Events beyond law. Events so numerous and so uncoordinated that flaunting their freedom from formula, they yet fabricate firm form." (3) These events, not of some new kind, but the elementary act of question to nature and a probability-guided answer given by nature, the familiar everyday elementary quantum act of observer-participancy. (4) Billions upon billions of such acts giving rise, via an overpowering statistics, to the regularities of physical law and to the appearance of a continuous spacetime. Far though one is from seeing how to spell out this vision, let alone appraise it, this is one conception of what it would mean to "understand geometry in terms of pregeometry."

In brief, we confront two imperatives and one great issue. First, the gates of time tell us that physics must be built from a foundation that has no physics; or still more briefly: "Must Build." Second, elementary quantum acts of observer-participatorship: "Do Build." Finally, how are billions upon billions of these elementary building acts organized-- if they are--to make up the grand structure that we call "reality"; or, in brief: "How Build?" No more attractive

clue offers itself for attacking this great issue than the
way information is processed to make "meaning." On what else
can a comprehensible universe be built but on the demand for
comprehensibility?

Getting a small but significant part of quantum mechanics
out of the demand that bits of information should produce
comprehensibility is the achievement of the following paper
by W.K. Wooters.


ACKNOWLEDGMENT

Appreciation is expressed to Adrienne L. Harding for help
with the literature.


REFERENCES

Cohen, P. (1966). <u>Set</u> <u>Theory</u> <u>and</u> <u>the</u> <u>Continuum</u> <u>Hypothesis</u>,
W.A. Benjamin, New York.

Ehrenfest, P.A. (1917). "In what ways does it become manifest in the fundamental laws of physics that space has
three dimensions?", in P.A. Ehrenfest, <u>Collected</u> <u>Scien-</u>
<u>tific</u> <u>Papers</u>, North-Holland Publishing Co., Amsterdam
1959; the laws of gravitation and planetary motion, the
dualism of "translation-rotation, force-pair of forces,
electric field-magnetic field," and the "integrals of the
equations of vibration" make sense only in three-
dimensional space.

Feynman, R.P. (1942). <u>The</u> <u>Principle</u> <u>of</u> <u>Least</u> <u>Action</u> <u>in</u>
<u>Quantum</u> <u>Mechanics</u>, doctoral dissertation, Princeton
University.

Feynman, R.P. and Hibbs, A.R. (1965).  Quantum Mechanics and
    Path Integrals, McGraw-Hill, New York.

Gödel, K. (1931).  "Über formal unentscheidbare Sätze der
    Principia Mathematica und verwardte Systeme I," Monatsch.
    Math. Phys. 38, 173-198, English translation by B.
    Meltzer, On Formally Undecidable Propositions, with an
    introduction by R.O. Braithwaite, Basic Books, New York,
    1962; outlined and discussed in E. Nagel and J.R. Newman,
    Gödel's Proof, New York University Press, New York, 1958.

Gödel, K. (1934).  "On undecidable propositions of formal
    mathematical systems," in M. Davis, ed., The Undecidable,
    Raven Press, Hewlett, New York, 1965; translation of
    mimeographed notes of lectures given in 1934.

Heisenberg, W. (1930).  "Die Selbstenergie des Elektrons,"
    Zeitschrift für Physik 65, 4-13.

Kadyshevskii, V.G. (1963a).  "Various parametrizations in
    quantized spacetime theory," Soviet Physics-Doklady 7,
    1031-1036.

Kadyshevskii, V.G. (1963b).  "A model of scalar field theory
    in quantum space-time," Soviet Physics-Doklady 7, 1138-
    1141.

Kant, I. (1868).  Kritik der reinen Vernunft, English trans-
    lation by F.M. Müller, Critique of Pure Reason, Anchor
    Books, Garden City, New York, 1966, page 24; three dimen-
    sions of space and other laws of nature as a "precondition
    for the possibility of phenomena"; see however chapter 11
    in A. Grünbaum, Philosophical Problems of Space and Time,
    Alfred A. Knopf, New York, 1963.

Kogut, J.B. (1978).  "A review of developments in lattice
gauge theory:  extreme environment and duality transforma-
tions," panel discussion, in J.E. Lannutti and P.K.
Williams, Current Trends in the Theory of Fields,
Tallahassee - 1978, A Symposium in Honor of P.A.M. Dirac,
AIP Conference Proceedings No. 48, Particles and Fields
Subseries No. 15, American Institute of Physics, New York,
1978.

Maxwell, C.M. (1877).  "On approximate multiple integration
between limits by summation," Proceedings of the Cambridge
Philosophical Society 3, 39-40; see also his comment, "In
cases of continuous stress...analytical methods may be
explained, illustrated, and extended by considerations
derived from the graphic method (of 'reticulation')" at
the end of his paper, "On reciprocal figures, frames, and
diagrams of forces," Transactions of Royal Society of
Edinburgh 26; both papers are reprinted in W.D. Niven,
ed., The Scientific Papers of James Clerk Maxwell, Volume
II, Dover Publications, New York, 1952, pages (603-607),
161-207.

Misner, C.W., Thorne, K.S., and Wheeler, J.A. (1973). Gravitation, W.H. Freeman and Co., San Francisco.

Patton, C.M. and Wheeler, J.A. (1975).  "Is physics legislated
by cosmogony?," in C.J. Isham, R. Penrose, and D.W.
Sciama, eds., Quantum Gravity, Clarendon Press, Oxford;
reprinted in R. Duncan and M. Weston Smith, eds.,
Encyclopedia of Ignorance, Pergamon Press, 1977, pages
19-35.

Peirce, C.S. (1940). The Philosophy of Peirce: Selected
    Writings, ed. by J. Buchler, Routledge and Kegan Paul,
    London; paperback reprint under the title Philosophical
    Writings of Peirce, Dover, New York, 1955, p. 358.

Poincaré, H. (1913). Dernières Penseès, English translation
    by J.W. Bolduc, Mathematics and Science: Last Essays,
    Dover Publications, New York, 1963, Ch. 3; pp. 27-28.
    Space as analyzed, not metrically, but via analysis situs
    (topology in the large), shows itself to be three-
    dimensional.

Regge, T. (1961). "General relativity without coordinates,"
    Nuovo Cimento 19, pp. 558-571.

Reichenbach, H. (1929). Philosophie der Raum-Zeit-Lehre, W.
    de Gruyter and Co., Berlin; English translation by M.
    Reichenbach and J. Freund, The Philosophy of Space and
    Time, Dover Publications, New York, 1958, Chapter 44,
    pp. 275-279; the dimensionality of space is well-defined,
    not being changed by any change in parametrization.

Snyder, H. (1947a). "Quantized space-time," Physical Review
    71, 38-41.

Snyder, H. (1947b). "The electromagnetic field in space-
    time," Physical Review 72, 68-71; see also Kadyshevskii
    (1963a, 1963b).

Weyl, H. (1963). Philosophy of Mathematics and Natural
    Science, Atheneum, New York, p. 36; Weyl notes that only
    in a space with an odd number of dimensions "will darkness
    follow the extinction of a candle," gauge invariance holds
    only for three dimensions, and refers to other considera-
    tions and others who have asked "why three dimensions?".

Wheeler, J.A. (1964). "Geometrodynamics and the issue of the final state," in C.M. DeWitt and B.D. DeWitt, _Relativity, Groups, and Topology_, Gordon and Breach, New York.

Wheeler, J.A. (1971). Note book entry, "Pregeometry and the calculus of propositions," 9:10 a.m., April 10; seminar, Department of Mathematics, Kings College, London, May 10; Letter to L. Thomas, "Pregeometry and Propositions," June 11, unpublished.

Wheeler, J.A. (1978). "The 'past' and the 'delayed-choice' double slit experiment," in A.R. Marlow, ed., _Mathematical Foundations of Quantum Theory_, Academic Press, New York.

Wheeler, J.A. (1979). _Frontiers of Time_, North-Holland, Amsterdam; also in N. Toraldo di Francia and B. van Fraassen, eds., _Rendiconti della Scuola Internazionale di Fisica_, "Enrico Fermi," LXXII Corso, _Problems in the Foundations of Physics_, North-Holland, Amsterdam.

# INFORMATION IS MAXIMIZED IN PHOTON POLARIZATION MEASUREMENTS

W. K. Wootters

Center for Statistical Mechanics
and
Center for Theoretical Physics
The University of Texas at Austin
Austin, Texas

In this paper we will consider the following question: Does a plane polarized ensemble of photons transfer, upon being measured, as much information about its plane of polarization as it conceivably could, given that the measurement of each photon has just two possible outcomes, and that the polarization is expressed only in the probabilities of the two outcomes?  We shall find that the answer is "yes"--the information is transmitted in the best way.

The reason this question arises--and it arises not just for photons but for any quantum system--is that in quantum mechanics there are more ways to prepare a system than there are outcomes of a measurement on that system.  In our example, the photons may be prepared by letting a beam of light pass through a polarizing filter.  There is a continuum of different preparations, one for each orientation of the filter (i.e., a continuum of possible planes of polarization), and yet any given orientation of the analyzing device gives just two possible outcomes, "yes" (partial evidence of agreement between the orientations of the polarizer and analyzer) and



Copyright © 1980 by Academic Press, Inc.
All rights of reproduction in any form reserved.
ISBN 0-12-473260-7

"no" (partial evidence against their agreement).  It is therefore impossible for a photon's preparation to be completely expressed in the outcome of a measurement, simply because the value of a continuous variable (the angle defining the orientation of the filter) cannot be expressed in a single binary decision.  Even an ensemble of N photons cannot transmit the information perfectly.  Our question is:  Do they transmit as much as they could?

Suppose that each photon is measured by letting it pass through a Nicol prism, from which it emerges in one of two possible directions, corresponding to the two polarizations "vertical" and "horizontal".  Thus an ensemble of N photons will yield a sequence of vertical and horizontal outcomes, for example, VVHVHVVVH...  An observer studying this sequence will learn something about the orientation of the filter (we suppose he knew nothing about it initially) by noting the relative frequencies of occurrence of V and H.  He uses the fact that the probability of the vertical outcome is $p_V = \cos^2\theta$, where $\theta$ is the angle between the preferred axis of the polaroid filter and the vertical axis.  (We will always take $\theta$ to be the smaller of the two angles between these axes; i.e., $0 \leq \theta \leq \frac{\pi}{2}$.)  If, for example, there are very few H's in the sequence, the observer knows that $\theta$ is close to zero.

This is how the N photons transmit information.  Is it the best <u>conceivable</u> way of transmitting information in N binary symbols?  It is clear that the answer to this question, as it now stands, is "no".  The reason is that the best conceivable way of transmitting information would certainly make use of the <u>order</u> of the symbols.  For example, one could send the first N digits of the binary expansion of $\theta$.  These would

contain much more information than our observer actually gets

from N photons, since he uses only the relative frequencies

of the two outcomes.

In order to rule out this kind of coding, we now impose

our second restriction (the first being the existence of just

two possible outcomes), namely, that the value of $\theta$ can be

encoded only in the probabilities of the two outcomes.  Given

these two restrictions, we now wish to show that the proba-

bility law which the photons obey, $p_V(\theta) = \cos^2\theta$, is the best

possible law for transmitting the value of $\theta$.  We will assume

that the a priori probability measure for $\theta$ is proportional

to Lebesgue measure.

It is clear that the problem we have to solve can be

stated as an abstract communication problem, not involving

photons at all:  We imagine one person, who plays the role of

the photons, trying to communicate to a second person the

value of a continuous variable $\theta \in \left[0,\frac{\pi}{2}\right]$ by sending him N

binary symbols (let us say they can take the values zero and

one--these values are the analogs of the two outcomes "verti-

cal" and "horizontal").  Like the ensemble of photons, the

sender is required to encode the value of $\theta$ only in the prob-

abilities of the two different symbols; once these probabili-

ties are fixed he must throw dice or use some similar device

to decide whether any given symbol will be "zero" or "one".

He and the receiver have to agree upon a way of encoding $\theta$ in

the probabilities of "zero" and "one". This amounts to speci-

fying the probability of a "zero", $p_0$, as a function of $\theta$.

The problem is to find out what code they should choose in

order to transmit the value of $\theta$ most effectively in N symbols.

It should be emphasized that this problem requires no input from physics. But once we solve it, we can compare the solution to what actually happens in the photon polarization measurement.

It happens that the best choice of a code depends on N. That is, if the participants know in advance that they will send exactly one hundred symbols, the code they should choose is slightly different from the one they should choose if they are going to send one thousand symbols. What we will find is the limit of the best code as N goes to infinity. It will turn out that this limit is the function $p_0(\theta) = \cos^2\theta$. This is our main result. It shows that a large ensemble of photons transmits the value of $\theta$ in the best way, since the probability of one of the two outcomes of the polarization measurement is given by this same function, $p(\theta) = \cos^2\theta$.

Let us now demonstrate that $p_0(\theta) = \cos^2\theta$ is indeed the best code. First we need to see why the value of $\theta$ cannot be communicated exactly. The reason is that when the receiver has received $n_0$ zeros and $n_1$ ones ($n_0 + n_1 = N$), he cannot be sure that the probability with which the sender is transmitting zeros is exactly $\frac{n_0}{N}$. There is a slight uncertainty, $\Delta p_0$, in the probability, due purely to statistics, and it is given by (1)

$$\Delta p_0 = \left[\frac{\frac{n_0}{N}(1 - \frac{n_0}{N})}{N}\right]^{\frac{1}{2}} \cong \left[\frac{p_0(1 - p_0)}{N}\right]^{\frac{1}{2}} \tag{1}$$

This particular dependence of $\Delta p_0$ on $p_0$ and N follows from the fact that the N trials (i.e., the N events in which the sender decides between "zero" and "one") are independent. The existence of this statistical uncertainty is the basis of the following calculation. The "best" code is best precisely because it minimizes the "effect" of this uncertainty.

The receiver now translates his value of $\frac{n_0}{N}$, i.e., his estimate of $p_0$, into a value for $\theta$, using the code $p_0(\theta)$. Let us assume that the function $p_0$ is invertible and differentiable. Then the uncertainty in $p_0$ leads to an uncertainty in $\theta$ given by

$$\Delta\theta = \left|\frac{dp_0}{d\theta}\right|^{-1} \Delta p_0 = \left|\frac{dp_0}{d\theta}\right|^{-1} \left[\frac{p_0(1 - p_0)}{N}\right]^{\frac{1}{2}} \tag{2}$$

In both Eqs. (1) and (2) we have used the fact that N is large, so that the uncertainties are small.

At this point there are several ways we could choose to define the expression "best code". Let us first adopt the most straightforward approach, and say that a function $p_0$ is best if it minimizes the average uncertainty $\Delta\theta$. The reason the average must be taken is that, according to Eq. (2), the uncertainty in $\theta$ may actually depend on the value of $\theta$ which is being sent, since the quantities $p_0$ and $\frac{dp_0}{d\theta}$ on the right-hand side depend on $\theta$. We thus want to minimize the quantity

$$\langle\Delta\theta\rangle = \int_0^{\pi/2} d\theta \, (\Delta\theta) = \int_0^{\pi/2} d\theta \left|\frac{dp_0}{d\theta}\right|^{-1} \left[\frac{p_0(1 - p_0)}{N}\right]^{\frac{1}{2}}$$

The variational calculation is easiest if we rewrite $\langle\Delta\theta\rangle$ as an integral over the probability rather than as an integral over $\theta$. Thus,

$$\langle \Delta\theta \rangle = \int_0^1 d\xi \, J^2(\xi) \left[ \frac{\xi(1 - \xi)}{N} \right]^{\frac{1}{2}}$$

where $J(\xi) = \left| \dfrac{dp_0}{d\theta} \right|^{-1}$ evaluated where $p_0 = \xi$. Since $\langle \Delta\theta \rangle$ depends on the function $p_0$ only through $J$, we can take the variation of $\langle \Delta\theta \rangle$ with respect to $J$. The only restriction on $J$ is that

$$\int_0^1 d\xi \, J(\xi) = \frac{\pi}{2} \tag{3}$$

and this we can take into account by means of a Lagrange multiplier $\lambda$. We will therefore take the variation of the quantity

$$B = \frac{1}{\sqrt{N}} \int_0^1 d\xi \left[ J^2(\xi) [\xi(1 - \xi)]^{\frac{1}{2}} - \lambda J(\xi) \right].$$

This gives us

$$\delta B = \frac{1}{\sqrt{N}} \int_0^1 d\xi \left[ 2J(\xi) [\xi(1 - \xi)]^{\frac{1}{2}} - \lambda \right] \delta J(\xi) ,$$

which vanishes for every $\delta J$ if and only if

$$J(\xi) = \frac{\lambda}{2} [\xi(1 - \xi)]^{\frac{1}{2}} . \tag{4}$$

The value of $\lambda$ is determined by Eq. (3), and one finds that $\lambda = 1$. According to Eq. (4) and the definition of $J$, a function $p_0$ extremizes $\langle \Delta\theta \rangle$ if it satisfies

$$\left| \frac{dp_0}{d\theta} \right| = 2p_0^{\frac{1}{2}} (1 - p_0)^{\frac{1}{2}} .$$

The only invertible solutions are $p_0(\theta) = \cos^2\theta$ and $p_0(\theta) = \sin^2\theta$. The form of these functions is due in part to the fact that we restricted $\theta$ to the interval $\left[0, \frac{\pi}{2}\right]$. If, in our hypothetical communication problem, we had chosen to let the domain of $\theta$ be $[\theta_0, \theta_0 + \pi/2\gamma]$, we would have obtained as the best codes the functions $p_0(\theta) = \cos^2\gamma(\theta - \theta_0)$ and $p_0(\theta) = \sin^2\gamma(\theta - \theta_0)$. We have chosen the domain of $\theta$ in such a way that we can compare our result to the probability law obeyed by photons. If we had taken the domain to be $[0,\pi]$ or $[0,\frac{\pi}{4}]$, we would have ended up with a law appropriate for electrons or gravitons, respectively.

To show that our solutions minimize, and not merely extremize, $<\Delta\theta>$, let us take the second variation:

$$\delta^2 B = \frac{1}{\sqrt{N}} \int_0^1 d\xi \, 2[\xi(1 - \xi)]^{\frac{1}{2}} (\delta J)^2 > 0.$$

The second variation is always positive; so our extremum is in fact an absolute, and not just local, minimum.

The preceding calculation was based on the definition of a best code as one which minimizes $<\Delta\theta>$. One might wonder whether the answer we got--that the function $p_0(\theta) = \cos^2\theta$ is best--has any absolute significance or whether it is peculiar to our particular definition of "best". Another reasonable approach would be to require that the best code maximize the amount of <u>information</u> the receiver can expect to gain about $\theta$, in Shannon's sense of the word (2,3). This concept of the expected increase in information is well defined once one specifies an <u>a priori</u> probability measure on the interval $[0,\frac{\pi}{2}]$ (4). If we choose Lebesgue measure, we find that the code $p_0(\theta) = \cos^2\theta$ does in fact maximize the information (5). Thus the optimal code does not seem to depend critically on what we have chosen to extremize. It does depend, however, on choosing Lebesgue measure as the <u>a priori</u> probability measure. If, for example, the receiver already knows that $\theta$ is between $0$ and $\frac{\pi}{4}$, then $p_0(\theta) = \cos^2\theta$ is not the best code.

Formulating the problem in terms of Shannon's information has one advantage over the "$\langle\Delta\theta\rangle$ approach", namely, that we need not restrict the domain of $\theta$ to the interval $\left[0,\frac{\pi}{2}\right]$. Indeed, one could object to our considering only these values of $\theta$, when the set of all plane polarizations would be more naturally labelled by $\theta \in \left[0,\pi\right]$. The function $p_0(\theta) = \cos^2\theta$ turns out to be a best code--in the sense of maximizing information--even if we allow $\theta$ to range over this larger interval (5).

We conclude that our sender and receiver would do best to choose the code $p_0(\theta) = \cos^2\theta$. The fact that the photons operate according to the same probability law shows that they too transmit information in the best way. In short, photons don't waste words.

## Discussion

Another way of formulating the problem is illustrated in Figure 1. There one wants to maximize the number of equally spaced distinguishable "messages" (i.e., distinguishable values of $\theta$) in the interval $\left[0,\frac{\pi}{2}\right]$. Again this maximization is accomplished by the probability function $p(\theta) = \cos^2\theta$.

It is interesting that we arrived at the formula $p_0(\theta) = \cos^2\theta$ not by squaring a probability amplitude, but by starting with statistical considerations which gave us the uncertainty $\Delta p_0$ (Eq. (1)). Let us now try to find a simple explanation of how this has happened. One might at first have thought that the best code would be the linear function

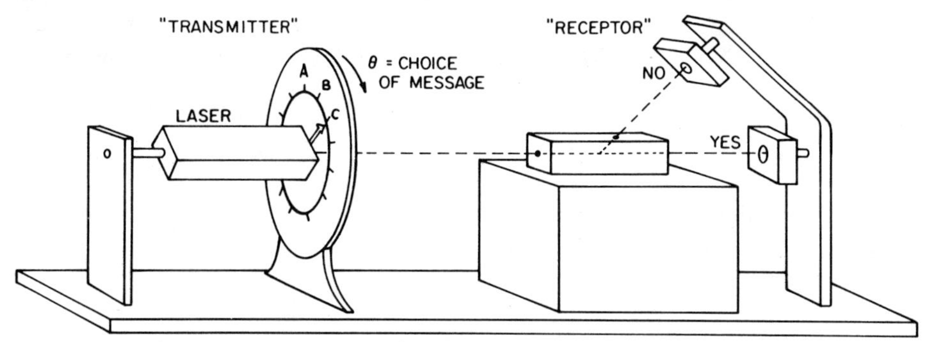


FIGURE 1.  Using stupid photons to transmit information. Warning of the direction (A, or B, or . . ., all equally spaced) of approach of danger is to be transmitted by properly orienting the laser's plane of polarization, opening its gate (not shown), and letting the laser give off its stock of 1000 photons with all the randomness of the law of radioactive decay (no timing, no Morse code, no controllability). How should the probability of "yes" depend on $\theta$ in order to maximize the number of distinguishable messages between $\theta_0$ and $\theta_0 + \pi/2\gamma$?  Answer (see text):  $\cos^2\gamma(\theta - \theta_0)$.  ($\gamma = \frac{1}{2}$ for electrons, 1 for photons, and 2 for gravitons.)

$p_0(\theta) = 1 - 2\theta/\pi$.  This code seems to treat all values of $\theta$ fairly, since it has the same slope everywhere, and since the slope is important in determining the uncertainty $\Delta\theta$.  (A steep slope makes $\Delta\theta$ small, because in that case a wide range of values of $p_0$ can correspond to a small range of $\theta$.)  Indeed, this linear code would be best <u>if</u> $\Delta p_0$ were a constant, independent of $\theta$.  But this is not the case, as we see from Eq. (1).  The uncertainty $\Delta p_0$ is greater when $p_0$ is near $\frac{1}{2}$ than when it is near 0 or 1.  In order to compensate for this greater uncertainty, the slope $\left|\dfrac{dp_0}{d\theta}\right|$ must be greater near the middle of the range of $\theta$.  This is why we get, instead of the linear function, the $\cos^2$ function, whose slope is steep around $\theta = \dfrac{\pi}{4}$, and which levels off near the endpoints of

$\left[0, \frac{\pi}{2}\right]$. In fact, this code is such that the uncertainty in $\theta$ is independent of $\theta$.

Can this result be generalized? This question actually contains two questions. First (see concluding paragraphs), is the efficiency which the photons exhibit a special case of a general principle of maximum information transfer in quantum measurements? Second, can we generalize our abstract communication problem? Yes, we can: the sender can be allowed more than just the two different symbols, zero and one--or "yes" and "no"--to transmit information. Let us say he can use M different symbols, $A_1, \ldots, A_M$. Again, we require that he encode his message (we will say shortly what this message might be) only in the probabilities of the M different symbols. What is the best way to do this encoding? As before, this question has nothing directly to do with physics. Nevertheless, it will be helpful for us to know the answer; we can use it in discussing quantum measurements with M possible outcomes. Therefore let us briefly consider this M-symbol communication problem.

In the case we considered above, where M = 2, the receiver was able to learn something about $\theta$ by noting how many zeros and ones were sent, and by estimating the probability with which the sender was sending each of these two symbols. In fact only one of these two probabilities is independent, say $p_0$, the other being $1 - p_0$. Thus the receiver had this one number, $p_0$, from which to deduce the value of the one variable $\theta$, using the agreed-upon code. In the general case, where there are M different symbols, the receiver will obtain estimates of M-1 numbers, $p_1, \ldots,$

$p_{M-1}$, where $p_i$ is the probability with which the sender is sending the symbol $A_i$. ($p_M$ is not independent of the other $p_i$'s, since $\Sigma p_i = 1$.) Therefore the sender can hope to transmit the values of M-1 continuous variables, one variable for each $p_i$. In other words, the set of possible "messages", instead of being just the interval $\left[0, \frac{\pi}{2}\right]$, can now be an (M-1)-dimensional set.

Let us assume that the sender and receiver are given such a set, D, of possible messages, D being a compact region of $R^{M-1}$. D comes equipped with an a priori probability measure, with respect to which "information" can be defined. The participants now have to decide on a way of assigning to each point in D, i.e., to each message, a set of probabilities $(p_1, \ldots, p_M)$. These are the probabilities with which the sender will send the various symbols $A_i$ when the message to be communicated is that particular point in D. This assignment of probabilities constitutes the code. As before, the receiver will be able to estimate these probabilities by counting how many of each different symbol $A_i$ he receives. But again, because of statistical fluctuations, he will not be able to deduce these probabilities exactly. The object is to choose the code in such a way that this uncertainty causes the receiver to lose as little information as possible regarding what message is being sent.

We will simply state without proof the solution to this problem:  The best way to encode the messages in the probabilities of the M different symbols is first to map the set D onto the unit sphere in $R^M$ in a measure-preserving way, and then to let the probabilities be the squares of the

Cartesian coordinates. That is, if x is a message (i.e., $x \in D$), it will be mapped into a point y on the unit sphere in $R^M$ with coordinates $(y_1, \ldots, y_M)$; when the sender has to transmit the message x, he will send the symbol $A_i$ with probability $p_i = y_i^2$. In this way the receiver will get the most possible information, on the average, regarding what message is being sent. Of course there are many different measure-preserving mappings of D onto the sphere (we are assuming the usual measure on the sphere, normalized so that the measure of the whole sphere is one), and each one gives a different code. But the point is that any code that can be obtained in this way is a "best" code, in the sense that the average information the receiver obtains is maximized; any other code does not allow the receiver to get as much information. Thus, there really is something special about letting the probabilities be the squares of "amplitudes".

This result indicates that there is a connection between statistical fluctuations in a sequence of independent trials, and the representation of probabilities as the squares of Cartesian coordinates of points on the unit sphere. The same connection has been used by Fisher and by Cavalli and Conterio in analyzing certain problems in genetics (6-8).

All of this suggests an interesting question: Is it possible to show that the quantum mechanical rule "probability = $|\text{amplitude}|^2$" follows from the requirement of maximum information transfer, or from some similar requirement? Such a requirement would presuppose the fact that there are more preparations than outcomes. Otherwise there is no advantage in a probabilistic code (cf. (9)). The idea of probability

as the square of a <u>real</u> amplitude does seem to arise naturally from these considerations.

So far we have no evidence in favor of complex amplitudes.
The restriction to real amplitudes in our photon polarization
experiment appears in the fact that the elliptical polarizations were not included as possibilities (they can be regarded as the complex linear combinations of the vertical
and horizontal polarizations).  Indeed, if a piece of bire-
fringent material were placed between the polarizing filter
and the Nicol prism, it could change the photon's state of
polarization (to elliptical) without affecting the outcomes
of the measurements.  In this sense there is some information
which is not transmitted at all, at least not in this experimental arrangement.  Work is currently being done to see
whether the problem can be changed in such a way that the
elliptical polarizations, and their analogs in the general
case of M possible outcomes, will fit into the scheme in a
natural way.

## ACKNOWLEDGMENTS

I would like to thank both Professor John A. Wheeler and
Dr. L. E. Reichl for many stimulating discussions and for
carefully reading the manuscript.

## REFERENCES


1.  Marek Fisz, <u>Probability Theory and Mathematical Statis-</u>
    <u>tics</u>, third edition, Wiley, New York 1963, section 5.2.
2.  C. E. Shannon, <u>Bell System Technical Journal</u>, <u>27</u>, 379
    (1948).

3. I. Csiszar, "Information Measures:  A Critical Survey," in _Transactions of the Seventh Prague Conference on Information Theory, Statistical Decision Functions, Random Processes_, Reidel, Boston 1978.

4. D. V. Lindley, Ann. Math. Statist. _27_, 986 (1956).

5. W. K. Wootters, in preparation.

6. R. A. Fisher, Proc. Roy. Soc. Edin., _42_, 321 (1922).

7. L. L. Cavalli-Sforza and F. Conterio, Atti Associazione Genetica Italiana _5_, 333 (1960).

8. Motoo Kimura, _Diffusion Models in Population Genetics_, Methuen, London 1964, pp. 23-25.

9. A. Landé, _New Foundations of Quantum Mechanics_, Cambridge University Press, London 1965, pp. 38-40.

# ROLES OF SPACE-TIME MODELS *

Carl H. Brans
Department of Physics
Loyola University
New Orleans, Louisiana

## 1. INTRODUCTION

Many physicists (including workers in general relativity) too often take for granted the central role that our space-time models, and the accompanying assumptions, play in the expression of physical theories. Thus, even though we all now know that the Euclidean axioms of a flat geometry are not experimentally valid when applied to real space, there are many other structures carried by our space-time models that may also not correspond to physical reality. It is the purpose of this talk to review briefly some of these questions.

To begin, we must note that space-time models play two distinct roles. First, as we learn from general relativity, space-time itself is a proper object for direct experimental investigation. This requires, of course, a careful specification of what experimental data are to correspond to what structures in the model, for example, light rays for straight lines. Second, space-time models are used as "scratch pads" on which computations are done, differential equations are expressed and solved, etc., to arrive at experimental predictions of theories not necessarily directly related to space-time questions. Of course, it might be noted that all experimental questions are ultimately answered in terms of space-time configurations and relationships of pointers of some sort. However, the cor-

---

*Supported in part by a grant from the Research Corporation.

27

Copyright © 1980 by Academic Press, Inc.
All rights of reproduction in any form reserved.
ISBN 0-12-473260-7

responding space-time structures are macroscopic, laboratory sized, for
which the familiar flat Minkowski structure is certainly valid.  It is the
scaling down of this structure that we would like to consider in this talk.

I am aware that many physicists believe that the investigation of
these questions is not likely to be rewarding, either because there are
now no pressing empirical data motivating their consideration, or because
the problems are simply too difficult and we do not have the required mathematical technology to solve them. While the latter objection may prove
to be correct, I certainly believe that the theoretical, if not experimental, difficulties associated with the inevitable singularities in general
relativity, the necessity for using infinite renormalization schemes in
field theories, and the observational limitation on the measurement of
space-time structures set by quantum theory all point to the basic inade-
quacy of our current smooth manifold models.

## II. SPACE-TIME AS A PHYSICAL OBJECT

The physical basis for geometry is evident from the origins of the
subject. The set of points defined by the earth's surface, or some physical plane surface, provides a definite arena for the testing of geometric
axioms.  The next, crucial, step is to extend these concepts to sets consisting of "empty" space points.  However, the absolute nature of the
points constituting the geometric set inherited from the study of sets of
physical particles was retained.  This is illustrated by the ether theories, for which the absolute point structure of space provides among other
things an absolute rest frame.  The ideas of Berkeley and Mach come to
mind here.

Einstein's insight into the unobservability of this absolute structure
of course led to special relativity. General relativity followed from the
questioning of the preferred nature of inertial coordinates and the flat
metric.[1] However, a brief look at a space-time diagram is enough to convince

us that the old space ether has been replaced with a new space-time one, in which the individual space-time events preserve their individuality for all observers in the same manner as space points do in an ether theory. This leads to the question:

> Is some generalization of the relativity principle
> possible (and useful) in which the absolute nature
> of the point set of space-time is replaced by a
> relative one?

We might also consider a less ambitious idea in which the identity of the point set structure is maintained but for which the topology and differentiability structures are not invariant. Thus, why should all observers organize space-time in the four-dimensional manifold that we do? A partial answer to this conjecture is suggested by the discussion of quantum logic and space-time below.

A direct approach to space-time as a physical object is provided by questioning what we might be missing by interpolating space-time macroscopic questions all the way down to points. The motivation for expecting that such interpolation might be unjustified is provided by the well known limitations on measurements demanded by quantum theory. A simple way of phrasing this idea might be to ask for the microscopic "gauge" inherent in interpolation from the macroscopic.

Let us approach space-time structure through the family of confinement questions. We can imagine idealized detectors and let questions, q, q', etc., refer to whether or not an event occurs in a corresponding detector (which of course includes time limits).

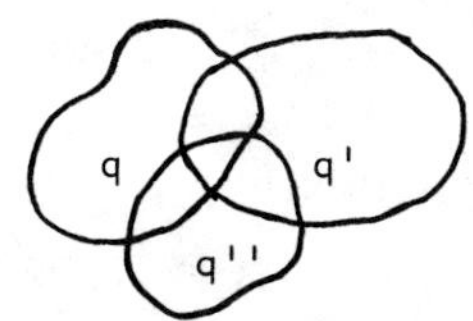

From macroscopic experience, we find that this family has the familiar partial ordering of a Boolean poset of subsets of a point set. For the diagram above, for example, $q \cap (q' \cup q'') = (q \cap q') \cup (q \cap q'')$. If we now use filters to construct the topology, we arrive at the familiar $R^4$. The question we must now consider is what effect the well known quantum limitations on measurement has on this process. To this end, we can divide the set of questions into macroscopic and microscopic, say M and m respectively, with no element of M preceding one of m. What is usually done in constructing space-time models along these lines is to interpolate the observed Boolean structure of M down to m. However, we have no observable motivation for doing so.  In fact, quantum theory strongly suggests that a non-Boolean structure for m is possible.  Thus we are led to the question:

What could we be missing by interpolating

a Boolean structure for m?

As a simple example, consider $R^4 \times S$, where S is some vector space. Let M be the set of questions referring to inclusion in V X S where V is some subset of $R^4$, while m refers to membership in spaces of the form V X P where P is a linear subspace of S and the ordering is the usual non-Boolean one for linear subspaces. The structure of M is thus non-Boolean, but we have no indication of this fact from the ordering of M, and completely miss it by interpolating down from M.

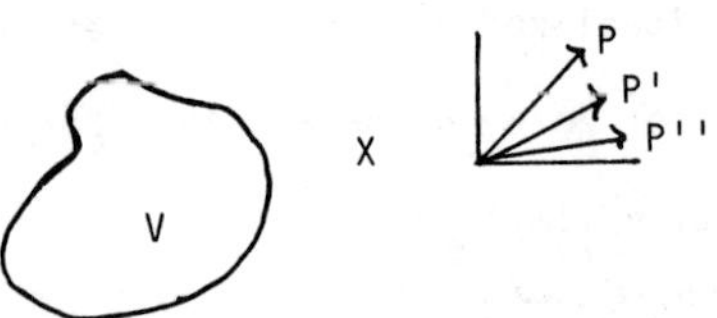


For this diagram, clearly $(V \times P) \cap ((V \times P') \cup (V \times P'')) \neq$

$((V \times P) \cap (V \times P')) \cup ((V \times P) \cap (V \times P''))$.

Marlow[2] has provided some interesting insights on the process by which such filters are used to construct topological spaces.

III. SPACE-TIME AS A "SCRATCH-PAD".

It is appropriate that quantum theory, which was so firmly rooted in operationalism, should have given rise to the development of quantum logic which epitomizes the operational spirit. Recall that the Hilbert space unification of the Schroedinger and Heisenberg formulations of quantum theory by Dirac and von Neumann was a great achievement in the early days of the theory. Later, von Neumann and others extracted from the Hilbert space formulation the operational essence which came to be known as quantum logic. In one formulation of this system, results of physical measurements are reduced to their binary representation so that all questions are simply yes/no ones. Quantum logic then studies families of such questions, together with states and a probability function from questions and states to the unit interval. This structure then has a natural partial ordering, which in general will be non-Boolean. Historically, such structures were first considered in the context of a Hilbert space in which both the binary questions and the (pure) states are one-dimensional projection operators, with the probability function given the familiar trace formula. However the poset of quantum logic is closer to the results of physical experimentation, so the question of the adequacy of Hilbert space to sustain an arbitrary structure of this sort is a natural one. Marlow[3] has answered this question affirmatively, although the representation may be redundant and not necessarily the simplest one.

At this point, we may ask where space-time structures enter. On the one hand, the set of space-time confinement questions discussed in the previous section could be imbedded in the Hilbert space structure. However, space-time also enters when function or distribution representations of the Hilbert space are used, and especially in the formulation of specific theories in the quantum language. Thus, some form of space acts as the domain for functions representing states or operators and a particular theory makes use of this functional representation to describe its version of reality, by requiring solutions of differential equations, etc. The question then arises:

> What is the relationship between observational
>
> space, as investigated by confinement questions
>
> and the domain space for functions in a theory?

As an example of alternative space representations, consider the harmonic oscillator problem whose expression in terms of creation and annihilation operators, a and a*, is described by the hamiltonian,

$$H = a^*a + 1/2,$$

or in terms of the usual spatial representation,

$$H = (-(d/dx)^2 + x^2)/2.$$

The space of states can be represented as a Hilbert space on which a is a ladder operator. This space can of course also be represented in terms of certain classes of functions over the real line, with $a = (d/dx+x)/\sqrt{2}$. In this representation, the eigenstates of H are represented by exponentially damped Hermite polynomials. However, there is a different "spatial" representation, in which the functions are much simpler. In fact, consider the Hilbert space of analytic functions of one complex variable, z, with inner product defined by integrations using $\exp(-z\bar{z})$ for measure. In this representation, a becomes merely d/dz while a* is multiplication by z, and the eigenfunctions of H are simply $z^n$. The relationship between z and the usual space variable x is $x = (z+d/dz)/2$. The point of this example is that the space for our functional representation is arbitrary, although this example has the obvious shortcoming that there is no correspondence limit relationship between C, the space for z, and R, the space for x, while it is the latter that accurately mirrors our macroscopic data on space. Nevertheless, this example does illustrate at least a partial extension of the relativity principle in the sense that while the theory can be expressed in either C or R in such a way as to produce identical operational predictions, the spaces C and R are clearly topologically distinct.

Finally, there is another intriguing, and simple way to construct space from quantum theory, based on the notion that the simple yes/no dichotomy can be represented by a two-dimensional Hilbert space, whose symmetry group is U(2), which is obviously related to SO(3,R). Thus we may be led to construct $R^3$ from a Hilbert space symmetry. Variations and extensions of this idea are discussed by von Weisaecker[4] and Finkelstein[5].

## IV. CONCLUSION

I have tried to provide a brief, cursory survey of some aspects of the roles and problems of space-time models in physics.  Certainly few answers were given.  However, anyone attempting to relate quantum theory and general relativity must reflect on the distinguished role of space-time, as well as the essential un-observability of the crucial features of our models in view of quantum limitations.  The history of both of these theories provides remarkable examples of the importance of questioning essentially non-testable hypotheses.

## REFERENCES.

1.    Trautmann, Reports on Math. Phys., 1, 29 (1970), has given an interesting account of the progression of relativity theories using bundle techniques.  He too suggests an extension of the relativity principle.

2.    A. R. Marlow,"Empirical Topology",to be published.

3.    A. R. Marlow, presentations to this conference.

4.    L. Castell, M. Drieschner, C.F. von Weizsacker,"Quantum Theory and the Structures of Time and Space", two volumes, Carl Hansen Verlag, Munchen, 1975 and 1977.

5.    D. Finkelstein, presentation to this conference.

# AN AXIOMATIC GENERAL RELATIVISTIC
# QUANTUM THEORY

A. R. Marlow[*]
Department of Physics
Loyola University
New Orleans, Louisiana  70118

## INTRODUCTION

An axiomatic model for a fully relativistic quantum theory is
developed in this paper.  The basic structure consists of four axioms
imposed on an operational quantum logical universe of discourse.  From
general embedding theorems recently established it is shown that the
standard computational framework of quantum theory exists for the
model.  While the resulting structure is general enough to deal with all
current physical theories in the usual context of infinite-dimensional
Hilbert space, what is most interesting from the present author's point
of view is that nontrivial and empirically more realistic models seem to
require finite-dimensional Hilbert space.  In such a context, of course,
the renormalization problems of standard relativistic quantum theory
do not arise.  Probably the most novel feature of any model of the proposed type, whether in finite or infinite dimensions, is the existence
of proper time operators representing fundamental observables in terms
of which everything else can be constructed.

---

* Work supported by a Research Corporation Grant.



Copyright © 1980 by Academic Press, Inc.
All rights of reproduction in any form reserved.
ISBN 0-12-473260-7

The paper is divided into four sections, with the fourth section containing more speculative proposals for further research. Section I gives the basic formal axiomatic and mathematical structure. For the reader who is unfamiliar with quantum logic or more interested in the motivation and interpretation of the formal structure, a prior reading of section II might be recommended. Section III then completes the more relativistic aspects of the structure in terms of a definition of chronology, the existence of proper time operators, and our last two axioms, which if left unweakened, result in a finite-dimensional Hilbert space model for the universe.

U followed by a numeral signifies an axiom for the universe. D a definition and T a theorem. □ designates the end of a formal statement. Our remaining notation is either fairly standard or defined as we go along. In particular $P(Q)$ is the set of all subsets of a set Q (power set), and $A^Q$ is the set of all mappings from Q into A.

I.  Basic Axiomatic Structure

> We choose for our universe of discourse


<u>U.</u>  A nonempty set $Q$ (physical questions) , a mapping $*: P(Q) \to Q$ (conjunction), and a function $P: Q \to [0,1]$ (probability).  $\square$


Such a triplet $U = \{Q, *, P\}$ will be called a <u>physical universe</u> if it satisfies the following two axioms:


<u>U1.</u>  $\forall\, q \in Q,\, * \{q\} = q.$  $\square$


<u>U2.</u>  $\forall\, R, S \in P(Q),\, R \subset S \Rightarrow P(*S) \le P(*R).$  $\square$


There are two reasons now for defining a functional representation of $U$:

1)  We want to pave the way for embedding $U$ in a convenient computational setting;

2)  we do not want to distinguish two physical questions unless the probability function $P$ gives grounds for such a distinction.

Both aims are accomplished by defining

<u>D1.</u>  $\forall\, R, S, T, \in P(Q),$

$$\tilde{R}(S) = P(*[R \cup S])$$

$$\tilde{R} * \tilde{T}(S) = [R \cup T]^{\sim}(S) = P(*[R \cup T \cup S]).$$

The function $\tilde{R} \in [0,1]^{P(Q)}$ given by $S \to \tilde{R}(S)$ will be called the function representation of $R$, and we define $P_{\sim}(R) = \{\tilde{M}: M \subset R\}$ and $R_{\sim} = \{\tilde{q} = \{q\}^{\sim}: q \in R\} \subset P_{\sim}(R)$.  The composition $*$ on $P_{\sim}(Q)$ defined by $(\tilde{R}, \tilde{T}) \to \tilde{R} * \tilde{T}$ will be called <u>conjunction on $P_{\sim}(Q)$</u>, or simply <u>conjunction</u> when no confusion with the earlier mapping on $P(Q)$ is likely.  $\square$


<u>T1.</u>  $P_{\sim}(Q)$ with the composition $*$ is a complete abelian indempotent semigroup with identity $\tilde{I} = \emptyset$ and null element $\tilde{Q}$.  $\square$

Proof: Since a complete semigroup is simply a set with a completely associative composition law (i.e., associativity holds even in products involving infinitely many elements), we need to verify that $*$ is abelian indempotent and completely associative. But these properties follow trivially from D1 and the corresponding properties of set theoretical unions. That the function representations of the empty set $\emptyset$ and the full set Q serve as identity and null element follows from $\emptyset \cup R = R$ and $Q \cup R = Q$. Q.E.D.


<u>T2.</u>   $\{\mathcal{P}_{\sim}(Q), \leq\}$ is a meet semilattice with $\tilde{R} * \tilde{T}$ as the meet (greatest lower bound) of elements $\tilde{R}$ and $\tilde{T}$, where $\leq$ is the natural partial ordering of real-valued functions. $\square$

Proof: Any abelian indempotent semigroup is a meet semilattice (by definition) if the partial ordering is defined by $\tilde{R} \leq \tilde{T} \equiv \tilde{R} = \tilde{R} * \tilde{T}$. To see that this ordering is the natural ordering of real functions we have: $\tilde{R} = \tilde{R} * \tilde{T} \Leftrightarrow P(*[R\cup S]) = P(*[R\cup T\cup S])$, $\forall S \in \mathcal{P}(Q)$, which implies (by U2) $P(*[R\cup S]) \leq P(*[T\cup S])$, $\forall S \in \mathcal{P}(Q)$, or $\tilde{R} \leq \tilde{T}$. Going in reverse order now, $P(*[R\cup S]) \leq P(*[T\cup S])$, $\forall S \in \mathcal{P}(Q)$, implies $P(*[R\cup V]) \leq P(*[T\cup R\cup V])$, $\forall V \in \mathcal{P}(Q)$ (by substitution of $R\cup V$ for S). But, by U2, $P(*[T\cup R\cup V]) \leq P(*[R\cup V])$, and so $P(*[R\cup V]) = P(*[R\cup T\cup V])$, $\forall V \in \mathcal{P}(Q)$, or $\tilde{R} \leq \tilde{T} \Leftrightarrow \tilde{R} = \tilde{R} * \tilde{T}$.          Q.E.D.


There are several types of universe U that we can classify now:

<u>D2.</u>   Call a universe U <u>empty</u> if P is the zero function in $[0,1]^Q$ Writing $\tilde{0}$ for the zero function in $[0,1]^{\mathcal{P}(Q)}$, the empty case can also obviously be characterized by $\mathcal{P}_{\sim}(Q) = \{\tilde{0}\}$. For nonempty universes we say

    a)  U is nonstatistical if $P(Q) \subset \{0,1\}$;

    b)  U is statistical if $P(Q) \not\subset \{0,1\}$.

From now on "universe" without further qualification will always mean "nonempty universe". $\square$


The purpose of the next few definitions is to introduce the notions necessary for embedding $\mathcal{P}_{\sim}(Q)$ into a bounded dual poset, since it is known that every such poset can be embedded into the computational structure of standard quantum theory in Hilbert space (Cf. Ref. 1, 2, for the relevant definitions and proofs).

__D3.__ For a universe U let $\widetilde{R}' \in [0,1]^{\mathcal{P}(Q)}$ be defined by $\widetilde{R}' = \widetilde{1} - \widetilde{R}$ for each $R \in \mathcal{P}(Q)$, and let $\mathcal{P}'(Q) = \{\widetilde{R}' : R \in \mathcal{P}(Q)\}$. Form now the union $\widetilde{\mathfrak{D}} = \mathcal{P}(Q) \cup \mathcal{P}'(Q)$, and allow the slight abuse of writing arbitrary elements of $\widetilde{\mathfrak{D}}$ as $\widetilde{R}$, $\widetilde{T}$, etc. Extend the mapping $\widetilde{R} \to \widetilde{R}'$ to all of $\widetilde{\mathfrak{D}}$ by, $\forall \widetilde{R} \in \widetilde{\mathfrak{D}}$, $\widetilde{R}' = 1 - \widetilde{R}$, and let $\leq$ mean the natural ordering of functions on $\widetilde{\mathfrak{D}}$. Finally, define the relation $\perp$ on $\widetilde{\mathfrak{D}}$ (<u>orthogonality</u>) by $\widetilde{R} \perp \widetilde{T} \equiv \widetilde{R} \leq \widetilde{T}'$. $\square$


__T3.__ For a nonempty universe U, the triplet $\{\widetilde{\mathfrak{D}}, \leq, '\}$ of <u>D3</u> is a bounded dual poset. $\square$

Proof: By a bounded dual poset is meant a poset $\mathfrak{D}$ with a greatest and least element (the "bounded" property) and a mapping $' : \mathfrak{D} \to \mathfrak{D}$ ("duality") such that,

    1) $R \neq R'$

    2) $R'' = R$

    3) $R \leq T \Rightarrow T' \leq R'$

All these requirements are trivially verified for $\{\widetilde{\mathfrak{D}}, \leq, '\}$ with the possible exception of 1) above, where we must check the two cases $\widetilde{R} \in \mathcal{P}(Q)$ and $\widetilde{R} \notin \mathcal{P}'(Q)$. However, since $\widetilde{R} \in \mathcal{P}'(Q) \Leftrightarrow \widetilde{R}' \in \mathcal{P}(Q)$, both cases can be reduced to showing that, for $\widetilde{R} \in \mathcal{P}(Q)$, $\widetilde{R} = \widetilde{R}'$ contradicts the assumption of a nonempty universe. We have, $\widetilde{R} \in \mathcal{P}(Q)$, $\widetilde{R} = \widetilde{R}' \Rightarrow$ $\widetilde{R} = \widetilde{1} - \widetilde{R} \Rightarrow 2\widetilde{R} = \widetilde{1} \Rightarrow \forall S \in \mathcal{P}(Q)$, $2P(*[R \cup S]) = P(*S) \Rightarrow \forall T \in \mathcal{P}(Q)$, $2P(*[R \cup T]) = P(*[R \cup T])$ (by substitution of $R \cup T$ for S) $\Rightarrow 2\widetilde{R} = \widetilde{R} \Rightarrow \widetilde{R} = \widetilde{0}$. But, by assumption, $2\widetilde{R} = \widetilde{1}$ and so, $\widetilde{R} \in \mathcal{P}(Q)$, $\widetilde{R} = \widetilde{R}' \Rightarrow \widetilde{1} = \widetilde{0} \Rightarrow P(*S) = 0$, $\forall S \in \mathcal{P}(Q) \Rightarrow$ U empty, and the theorem follows. Q.E.D.


__D4.__ For $S \in \mathcal{P}(Q)$ let $\hat{S} \in [0,1]^{\widetilde{\mathfrak{D}}}$ be the function defined by

$$\forall \widetilde{R} \in \mathfrak{D}, \hat{S}(\widetilde{R}) = \begin{cases} \dfrac{\widetilde{R}(S)}{\widetilde{1}(S)} & , \widetilde{1}(S) \neq 0 ; \\ 0 & , \widetilde{1}(S) = 0 ; \end{cases}$$

and let $\hat{\mathcal{S}} = \{\hat{S} : S \in \mathcal{P}(Q)\}$ (called the set of <u>physical</u> <u>states</u> on $\widetilde{\mathfrak{D}}$).

As in $\underline{D1}$, we introduce a composition $*$ on $\hat{S}$ by $\hat{S} * \hat{T} = (S \cup T)^{\wedge}$, so that $\{\hat{S}, *\}$ is a complete indempotent abelian semigroup with identity $\hat{1} = \hat{\emptyset}$ and null element $\hat{Q}$. In general, for $\mathfrak{Q}$ a bounded dual poset, an order-preserving function $\varphi \in [0,1]^{\mathfrak{Q}}$ (i.e., $\forall \, Q, R \in \mathfrak{Q}, \; Q \leq R \; \Rightarrow \; \varphi(Q) \leq \varphi(R)$) is called a $\underline{\text{measure}}$ on $\mathfrak{Q}$ if, $\forall \, Q \in \mathfrak{Q}$, $\varphi(Q') = \varphi(1) - \varphi(Q)$. A measure is called a $\underline{\text{probability}}$ $\underline{\text{measure}}$ if $\varphi(1) = 1$, and we write $\Phi_{\mathfrak{Q}}$ for the set of all measures on such a poset. A subset $\Psi \subset \Phi_{\mathfrak{Q}}$ is said to be full if $[\varphi(Q) \leq \varphi(R), \; \forall \, \varphi \in \Psi] \Rightarrow Q \leq R$. $\quad \square$


$\underline{T4.}$   For a nonempty universe, the set $\hat{S}$ of physical states is a full set
       of measures on $\tilde{\mathfrak{Q}}$. $\quad \square$

Proof: First note that, by D4, $\forall \, \tilde{R} \in \tilde{\mathfrak{Q}}, \; \tilde{R}(S) = \hat{S}(\tilde{R}) \, \tilde{1}(S), \; \forall \, S \in \mathcal{P}(Q)$ (since $\tilde{1}(S) = 0 \; \Rightarrow [\tilde{R}(S) = 0 \text{ and } \hat{S}(\tilde{R}) = 0]$). Then, $\forall \tilde{R}, \tilde{T} \in \tilde{\mathfrak{Q}}$,

$\tilde{R} \leq \tilde{T} \Leftrightarrow \forall \, S \in \mathcal{P}(Q), \; \tilde{R}(S) \leq \tilde{T}(S) \Leftrightarrow \forall \, S \in \mathcal{P}(Q), \; \hat{S}(\tilde{R}) \, \tilde{1}(S) \leq \hat{S}(\tilde{T}) \, \tilde{1}(S)$

$\Leftrightarrow \forall \, S \in \mathcal{P}(Q), \; \hat{S}(\tilde{R}) \leq \hat{S}(\tilde{T}).$ $\qquad$ Q.E.D.


$\underline{D5.}$   For $\mathfrak{Q}$ a bounded dual poset, a subset $\mathfrak{m} \subset \mathfrak{Q}$ is said to be a base
       for $\mathfrak{Q}$ if $1 \in \mathfrak{m}$ and, $\forall \, R \in \mathfrak{Q}, \; R \in \mathfrak{m} \Leftrightarrow R' \notin \mathfrak{m}$. $\quad \square$


$\underline{T5.}$   There exists a base for every bounded dual poset. $\quad \square$

Proof: For $\mathfrak{Q}$ the poset, consider the class $\overline{\mathfrak{Q}} \subset \mathcal{P}(\mathfrak{Q})$ of all subsets $\mathfrak{n} \subset \mathfrak{Q}$ such that $1 \in \mathfrak{n}$ and $R \in \mathfrak{n} \Rightarrow R' \notin \mathfrak{m}$. Let $\mathfrak{Q}$ be ordered by inclusion and consider a chain $\mathfrak{n}_{\alpha} \in \overline{\mathfrak{Q}}$. Then it is easy to check that $\cup_{\alpha} \mathfrak{n}_{\alpha} \in \overline{\mathfrak{Q}}$. Thus, since every chain in $\overline{\mathfrak{Q}}$ has an upper bound, there exists (by Zorn's Lemma) a maximal element $\mathfrak{m} \in \overline{\mathfrak{Q}}$, i.e. a maximal subset $\mathfrak{m} \subset \mathfrak{Q}$ such that $1 \in \mathfrak{m}$ and $R \in \mathfrak{m} \; \Rightarrow \; R' \notin \mathfrak{m}$. Then $\{R\} \cup \mathfrak{m} \in \overline{\mathfrak{Q}}$, which contradicts maximality of $\mathfrak{m}$. Hence, $\forall \, R \in \mathfrak{Q}, \; R \in \mathfrak{m} \Leftrightarrow R' \notin \mathfrak{m}$, and so $\mathfrak{m}$ is a base for $\mathfrak{Q}$. $\qquad$ Q.E.D.


$\underline{D6.}$   Unless otherwise specified, a Hilbert Space H will mean either
a real or complex Hilbert space, with $P_H$ the set of projection operators on H and $P_H^1$ the set of one-dimensional projections in $P_H$. $\quad \square$

The basic Hilbert space embedding theorem can now be given:

<u>T6.</u>  For a nonempty universe U, with $\widetilde{\mathfrak{D}}$ the bounded dual poset of D3 and $\hat{\mathfrak{S}}$ the full set of states of D4, let $\widetilde{\mathfrak{m}}$ be any base of $\widetilde{\mathfrak{D}}$. Then there exists a Hilbert space H and two mappings, $\pi: \widetilde{\mathfrak{D}} \to P_H$ and $\varphi: \hat{\mathfrak{S}} \to P_H^1$, such that, $\forall\ \widetilde{R},\ \widetilde{T} \in \widetilde{\mathfrak{D}};\ \widetilde{M},\ \widetilde{N} \in \widetilde{\mathfrak{m}};\ \hat{S},\ \hat{W} \in \hat{\mathfrak{S}}:$

1) $\pi(\widetilde{1}) = I \in P_H.$

2) $\pi(\widetilde{R}) = I - \pi(\widetilde{R}) = \pi(\widetilde{R})' \in P_H.$

3) $\hat{S}(\widetilde{R}) = \text{trace}(\pi(\widetilde{R})\ \varphi(\hat{S})).$

4) $\widetilde{M} \leq \widetilde{N} \Leftrightarrow \pi(\widetilde{M}) \leq \pi(\widetilde{N}).$

5) $\hat{S} \neq \hat{W} \Rightarrow \varphi(\hat{S})\ \varphi(\hat{W}) = 0.$

6) $\pi(\widetilde{R})\ \pi(\widetilde{T}) = \pi(\widetilde{T})\ \pi(\widetilde{R}).$ $\square$

Proof: Cf. Ref. 1, 2,  where the result is proved for any bounded dual poset and any set of measures.  Q.E.D.


Clearly the embedding in Hilbert space structure given by <u>T6</u> is overly restrictive by the usual standards of quantum theory, since

a) it embeds all states as pure states $[\varphi(\hat{S}) \subset P_H^1]$;

b) it embeds all states pairwise orthogonally $[\underline{T6},\ 5)]$;

c) $\widetilde{\mathfrak{D}}$ is embedded as an abelian set of projections $[\underline{T6},\ 6)]$.

However, <u>T6</u> does show what can be done, and we are now free to weaken our requirements as we choose in order to get a more compact and/or flexible mathematical model for physics.

Any quantum theoretical model worthy of the name will demand that conditions 1), 2) and 3) of <u>T6</u> be fulfilled, and since we want to keep as much of the natural ordering of physical questions as possible, we will also maintain requirement 4) on some base of $\widetilde{\mathfrak{D}}$. [Of course standard axiomatics at this point makes the <u>physical</u> assumption that the ordering on $\widetilde{\mathfrak{D}}$ is the ordering of projections on a Hilbert space, but since counter-examples (Cf. Ref. 2) seem to indicate the existence of possible physical situations where this cannot work, we prefer to stay as long as possible with weaker assumptions that can always be made to hold in our mathematical model without running the risk of physical difficulties -- i.e., we are still at the stage of constructing a mathematical language for physics. ]

We will generally dispense entirely with condition 6) of $\underline{T6}$ as a requirement for the quantum theoretical embeddings we construct, but we note that since it can be made to hold for all questions, we are guaranteed that we can include classical situations in quantum theory by requiring it to hold on suitably selected subsets of $\widetilde{\mathfrak{D}}$.

This still leaves various possibilities for weakening condition 5) of $\underline{T6}$ and modifying the range of the embedding of states so that we can deal conveniently with classical ensembles of states.

$\underline{D7.}$ For U a nonempty universe write $\widetilde{\mathfrak{D}}_U$ for the dual poset of $\underline{D3}$ and $\hat{\mathfrak{s}}_U$ for the set of states of $\underline{D4}$. Say that $\hat{\mathfrak{s}} \subset \hat{\mathfrak{s}}_U$ $\underline{generates}$ $\hat{\mathfrak{s}}_U$ if every element of $\hat{\mathfrak{s}}_U$ is a convex linear combination of elements in $\hat{\mathfrak{s}}$. For H a Hilbert space let $T_H^1$ be the set of non-negative trace 1 operators on H, and say that a subset $T \subset T_H^1$ $\underline{spans}$ a subset $T' \subset T_H^1$ if the range of every operator in $T'$ is contained in the subspace generated by the union of the ranges of the operators in T.

Given now a nonempty universe U and a base $\widetilde{\mathfrak{m}} \subset \widetilde{\mathfrak{D}}_U$ we say that a pair of mappings $\{\pi, \varphi\}$, $\pi: \widetilde{\mathfrak{D}}_U \to P_H$ and $\varphi: \hat{\mathfrak{s}}_U \to T_H^1$ is a $\underline{quantum}$ $\underline{mechanical}$ $\underline{embedding}$ of U ($\underline{QM}$ $\underline{embedding}$) with respect to $\widetilde{\mathfrak{m}}$ if 1), 2), 3) and 4) of $\underline{T6}$ are satisfied, together with:

5)    $\hat{S}, \hat{W} \in \hat{\mathfrak{s}}_U$, $\hat{S} * \hat{W} = 0 \Rightarrow \varphi(\hat{S})\varphi(\hat{W}) = 0$;

6)    $\varphi^{-1}(P_H^1) \subset \hat{\mathfrak{s}}_U$ generates $\hat{\mathfrak{s}}_U$, and there exists a subset

$\hat{\mathfrak{B}} \subset \varphi^{-1}(P_H^1)$, with $\varphi(S)\,\varphi(W) = 0$ for $\hat{S} \neq \hat{W}$, $\hat{S}, \hat{W}, \in \hat{\mathfrak{B}}$,

such that $\varphi(\hat{\mathfrak{B}})$ spans $\varphi(\hat{\mathfrak{s}}_U)$.

The set $\hat{\mathfrak{s}}_U^P \equiv \varphi^{-1}(P_H^1)$ will be called the $\underline{pure}$ $\underline{states}$ of the embedding; all other elements of $\hat{\mathfrak{s}}_U$ will be called $\underline{mixed}$ $\underline{states.}$    $\square$

$\underline{T7.}$ For a nonempty universe U and any base $\widetilde{\mathfrak{m}} \in \widetilde{\mathfrak{D}}_U$, there exists a quantum mechanical embedding of U with respect to $\widetilde{\mathfrak{m}}$.    $\square$

Proof: The embedding given by $\underline{T6}$ is a QM embedding with $\hat{\mathfrak{B}} = \hat{\mathfrak{s}}_U = \hat{\mathfrak{s}}_U^P = \varphi^{-1}(P_H^1)$.      Q.E.D.

Now on the general methodological principle that the more
mathematical structure a physical model possesses the more useful
it will be -- especially, in light of the history of physics, if the
structure is linear, algebraic, with Euclidean quadratic norm -- we
introduce the notion of an embedding in which the states are mapped
into $S_H^1$ , the unit sphere in the Hilbert space of Schmidt operators.
This space is also an algebra, and we will have the further advantage
that all states, whether pure or mixed, are represented by unit vectors
in $S_H^1$ .

**D8.**   For notational convenience we will write simply $\mathfrak{D}$, $\mathfrak{S}$ and $\mathfrak{B}$ for
the sets $\hat{\mathfrak{D}}_U$ , $\hat{\mathfrak{S}}_U$ and $\hat{\mathfrak{B}}$ in an embedding of a universe U, and we will
allow the omission of $\sim$ and $\wedge$ over the elements of these sets when no
confusion is likely; for embedding mappings such as $\pi$, $\varphi$ we often
write the arguments as subscripts (i.e., $\pi_R$ for $\pi(R)$ , etc.).

For a Hilbert space, let $S_H$ be the set of Schmidt operators on H,
i.e., $\alpha \in S_H \Leftrightarrow \alpha\alpha^* \in T_H^1$ . [Since $S_H$ is a Hilbert space under the
inner product $\langle \alpha, \beta \rangle$ = trace ( $\alpha^* \beta$) = trace ($\beta \alpha^*$) , $S_H^1$ is just the
unit sphere in $S_H$. It is also not difficult to check that $P_H^1$ =
$S_H^1 \cap T_H^1$ . ] For a subset $S \subset S_H$ write $\overline{S}$ for the Hilbert subspace
of $S_H$ generated by S.

A QM embedding will be said to be <u>second quantized</u> ($QM^2$) if, in
addition to $\pi$ and $\varphi$, a mapping $\eta : \mathfrak{S} \to S_H^1$ is specified such that

1) $\eta(\mathfrak{S}^P) \subset \overline{\eta(\mathfrak{B})}$   ;

2) $\forall S \in \mathfrak{S}, \eta_S \eta_S^* = \varphi_S$ .

An embedding of any type will be said to be coherent if $\hat{1}$ is
embedded as a pure state, i.e., if

3) $\hat{1} \in \varphi^{-1}(P_H^1) = \mathfrak{S}^P$.

Call $\hat{1}$ the universal state and say that $QM^2$ embedding is <u>universal</u> if

4) $\forall \alpha \in \eta(\mathfrak{B})$ , $\alpha \eta_1 = \alpha$.   $\square$

<u>T8.</u>    There exist coherent universal $QM^2$ embeddings for arbitrary

nonempty universes.    □

Proof: The embedding used in the proof of <u>T7</u> is already coherent and $QM^2$ if we choose $\eta = \varphi$, since $\mathcal{S} = \mathcal{S}^P = \mathcal{B}$, but to satisfy 4) above we have to choose $\eta$ in a different way. Start with the coherent QM embedding of <u>T7</u> and define $\eta_S = |S\rangle\langle 1|$, $S \in \mathcal{S}$, using Dirac dyadic notation with $|S\rangle$ a unit vector in the range of $\varphi_S$. This gives $\eta_1 = |1\rangle\langle 1| = \varphi_1$, and since $\mathcal{B} = \mathcal{S}$, $\overline{\eta(\mathcal{B})}$ consists of elements of the form $\alpha = \sum_{S \in \mathcal{S}} c_S \eta_S = \sum_{S \in \mathcal{S}} c_S |S\rangle\langle 1|$, so that $\alpha \eta_1 = \alpha$ is obviously satisfied along with conditions 1), 2), and 3) of D8.        Q.E.D.

In modern physics the tensor algebra of a Hilbert space is generally used to represent states regarded as composites of other states. To introduce this type of representation into our structure we have:

<u>T9.</u>    For U a universe and $\mathfrak{M} \subset \mathcal{Q}_U$ a base, there exists a coherent universal $QM^2$ embedding, with the tensor algebra $H^\otimes$ of a Hilbert space H as the Hilbert space of the embedding, such that, for all finite collection of states $S, T, V, W, \ldots \in \mathcal{S}$ with $S \neq 0$, and all questions $R \in \mathcal{Q}_U$,

$$S = T * V * W * \ldots \Rightarrow$$

$$\text{trace}\,(\pi_R \varphi_S) = \text{trace}\,(\pi_R \varphi_T \otimes \varphi_V \otimes \varphi_W \otimes \ldots).\quad □$$

Proof: We first note that $S = T * V * W * \ldots$ implies $S \leq T$, $S \leq V$, etc., when the states are considered as functions on $\mathcal{Q}_U$. Start now with the coherent universal $QM^2$ embedding of <u>T8</u>, which is based on the embedding given in <u>T6</u>. In the construction used to establish the latter embedding (Cf. reference 2) a separate Hilbert space $H_T$ is assigned to each state T, and a mapping $\pi^T$ of $\mathcal{Q}_U$ into the projections on $H_T$ is constructed such that, $\forall\, R \in \mathcal{Q}_U$, $T(R) = \text{trace}\,(\pi_R^T \varphi_T)$, where $\varphi_T$ is a fixed one-dimensional projection on $H_T$. The Hilbert space of the embedding is then given by $H = \sum_{T \in \mathcal{S}}^\otimes H_T$ and the projection corresponding to each $R \in \mathcal{Q}_U$ is defined by $\pi_R^H = \sum_{T \in \mathcal{S}}^\otimes \pi_R^T$. Moreover, $S \leq T \Rightarrow$

$\dim (H_S) \leq \dim (H_T)$. Hence there exist isometries $U_{S,T,V,W,\ldots}$ mapping $H_S$ into $H_R \otimes H_V \otimes H_W \otimes \ldots$ for each finite ordered set $(S,T,V,W,\ldots)$ of states with $S = T*V*W*\ldots \neq 0$, such that $U_{S,T,V,W,\ldots} \, \varphi_S \, U^*_{S,T,V,W,\ldots} = \varphi_T \otimes \varphi_V \otimes \varphi_W \otimes \ldots$, and using the axiom of choice, any selection of one such isometry for each ordered set yields an embedding satisfying the requirements of the present theorem if we define, $\forall \, R \in \mathbb{m}$, $\pi_R = \Sigma^{\oplus} U_{S,T,V,W,\ldots} \, \pi_R^S \, U^*_{S,T,V,W,\ldots}$, where the summation ranges over the ordered sets $(S,T,V,W,\ldots)$ such that $S = T*V*W*\ldots \neq 0$, and let $\pi_R'$, be the complementary projection $I - \pi_R$ on $H^{\otimes}$.    Q.E.D.

This completes most of the mathematical structure needed prior to the introduction of further physical axioms over and above U1 and U2. Observables more general than questions are introduced as question-valued measures (or in the context of a QM embedding, as projection valued measures, equivalently, self-adjoint operators) in the standard way (Cf. for example, reference 3).

II. <u>Interpretation and Motivation</u>

The interpretation used in this paper for the elements $q \in Q$ (the physical questions) is the, by now, fairly standard one of quantum logic: each q is regarded as a finite list of instructions for deciding unambiguously whether to record 1 (yes) or 0 (no) as the result or answer of the specific question q involved. Two lists q, r are considered distinct if their instructions differ in any way. Each individual instance of fulfillment of the instructions in a list q with consequent recording of a result 0 or 1 will be regarded as a separate performance of the question q.

Probability will be interpreted in the frequency sense, so that the value $P(q)$ of the universal probability P at a question q is understood as the ratio of the number of performances of q with result 1 to the total number of performances of q. We idealize only to the extent that we want P to reflect the total number of performances of q that will ever be made, while recognizing that all we really have to work with is the data that exists at any given epoch in the history of physics.

It is then the task of physical theory to formulate specific models, within the general language developed in the preceding section, that both incorporate all extant data while anticipating or predicting further results. The theorems of the preceding section simply guarantee the existence of models that incorporate all possible data, and so we at least have a sort of universal mathematical blackboard on which we can formulate and test specific models.

The main distinguishing feature of the present axiomatization of physical theory is the mapping $*$ (conjunction) from the set $\mathcal{P}(Q)$ of all subsets of questions into the set of questions itself, which, from the point of view of this paper, reflects the distinguishing feature of modern physics: a close operational scrutiny of conjunctive statements in general. The beginning of what is usually meant by modern physics can be traced to Einstein's operational analysis of temporal conjunctivity (simultaneity), and a good case can be made that the 1925-26 quantum revolution was simply the extension of that same spirit of operational analysis to all conjunctive statements ("observable A measured together with observable B").

In any event, our inclusion of the mapping $*$ as a primitive element of our structure declares that we are not willing to accept as meaningful the measurement of the conjunction of a set R of questions until a specific finite list of instructions $*R$ has been provided for carrying out the measurement.

Then U1 just posits the obvious requirement that the list $* \{q\}$ should be q itself, and, granted the frequency interpretation of probability, U2 guarantees that conjunction will live up to minimal expectations for such an operation, in the sense that every "yes" answer for a conjunction should count as a "yes" answer for each component of the conjunction.

We have already given the basic reasons for constructing a functional representation of our structure, but the precise method used warrants some further comment. Note that an expression such as $\tilde{R}$ (S) has no direct relation to either $*R$ or $*S$, but only to $*[R \cup S]$, the conjunction of all the questions in $R \cup S$, and a similar statement holds for the expression $\hat{S}(\tilde{R}) = \dfrac{\tilde{R}(S)}{\tilde{I}(S)} = \dfrac{P(*[R \cup S])}{P(S)}$ , i.e., the state

$\hat{S}$ evaluated on $\tilde{R}$. We could add a consistency axiom [(*)] that would en-
force a relation between R and *R, but we believe that this would
needlessly complicate the axiomatic structure and only serve to de-
emphasize the important point that a physical measurement depends on
the conjunction of all the factors present in the total situation. However,
this will not prevent us from giving the natural interpretation of con-
ditional probability to an expression of the form $\hat{S}(\tilde{R})$ , i.e., the
probability that every r $\in$ R has the answer yes, granted that we are
in a situation where every s $\in$ S has the answer yes, and we use the
further terminology given in the following definition:

<u>D10.</u>    If all the states in a given discussion are of the form W=T*S for
some S $\neq$ 0 , we refer to S as a <u>system,</u> and each T*S, T $\in$ $\mathcal{S}$, will be called
a <u>state of the system S.</u> Each function $\tilde{R} \in \tilde{\mathcal{Q}}$ of the form $\tilde{R}=\tilde{V}*\tilde{S}$, $\tilde{V} \in \tilde{\mathcal{Q}}$, will
be called a <u>question on the system S.</u> If a system is of the form $S = \prod_{i \in \Omega} *S_i$

(conjunction of a family $S_i$ , i $\in$ $\Omega$, where $\Omega$ is some labeling set), then
obviously every state W= T $*$ S is a state of each system $S_i$ separately,
and we will say that S is the system formed by the <u>interaction</u> of the
$S_i$ , i $\in$ $\Omega$, or simply the <u>composite</u> of the systems $S_i$ .  A family of

---

(*) One example would be: For all families $R_\alpha$ $\in$ $\mathcal{P}(Q)$ , $\alpha$ $\in \Omega$,

$P(*[\bigcup_{\alpha \in \Omega} R_\alpha]) = P(*[\bigcup_{\alpha \in \Omega} \{*R_\alpha\}])$ . Such an axiom simply

requires that the assignment of lists * R be chosen to be probabilistically
consistent with the assignment of lists by the $*$ operation on unions of
subsets. This axiom was implicit in the notation used in an earlier
paper along these lines, and was tacitly used in the proofs there, [4]
although erroneously omitted from the list of axioms presented (to be
supplied, it is hoped, by the charitable and sympathetic reader). In
the present context, no such axiom is necessary.

systems $\hat{S}_i$, $i \in \Omega$, will be said to be

     a) <u>independent</u> or <u>non-interacting</u> if

$$\forall \tilde{R} \in \tilde{\mathfrak{D}}, \quad [\,\overset{*}{\underset{i \in \Omega}{\Pi}} \hat{S}_i\,](\tilde{R}) = \underset{i \in \Omega}{\Pi} \hat{S}_i(\tilde{R}).$$

     b) <u>mutually exclusive</u> if

$$\forall\, i, j \in \Omega,\; i \neq j \;\Rightarrow\; \exists\, \tilde{R}_{ij} \in \tilde{\mathfrak{D}} : \hat{S}_i(\tilde{R}_{ij}) = 1,\; \hat{S}_j(\tilde{R}_{ij}) = 0.$$

Finally, define the <u>set of states on a system $\hat{S} \in \hat{\mathfrak{S}}$</u> by $\hat{\mathfrak{S}}_S = \{\, \hat{W} = \hat{T}*\hat{S},\ \hat{T} \in \hat{\mathfrak{S}}\,\}$, and the <u>set of questions on a system $\hat{S} \in \hat{\mathfrak{S}}$</u> by $\tilde{\mathfrak{D}}_S = \{\tilde{R} * \tilde{S} : R \in \mathcal{P}(Q)\,\} \cup \{\, [\tilde{R} * \tilde{S}]' \equiv (\tilde{1} - \tilde{R} * \tilde{S}) : R \in \mathcal{P}(Q)\,\}.$    $\square$

     Thus a system is to be regarded simply as a probability measure on the universe that is kept fixed as part of the background of a given discussion and defines a sort of sub-universe within the universe. There is no absolute distinction between states and systems, in the sense that what is a state in one context may be a system in another context. The motivation for associating interaction with conjunction is simply the intuitive notion that if several physical systems (i.e., physical situations, states of affairs) are actually present together, then all mutual interactions of which they are capable should be present and play a part in constituting the total state of affairs.

     Before any further discussion of interacting systems and their treatment in modern physics, some comments on mixed systems (or states) are in order. In the formalism developed so far, which states are mixed and which are pure has been a property of the embedding used rather than of the states themselves. This is necessitated by the fact that we can construct embeddings in which all states are embedded as one-dimensional projections, and hence as pure states in the usual quantum theoretical terminology. It is well known, however, that the distinction between pure and mixed states can be given solely in terms of the states themselves, without regard to any particular embedding: define a state $\hat{S}$ as physically mixed if it can be written as a countably convex combination of states $\hat{S}_i \neq \hat{S}$ (i.e., $\hat{S} = \underset{i \in \Omega}{\Sigma} \lambda_i \hat{S}_i$, $0 \leq \lambda_i < 1$,

and $\lambda_i \neq 0$ for an at most countable family $\hat{S}_i \neq \hat{S}$ ). The problem
comes in guaranteeing that there are enough physically pure states
(defined as states that are not physically mixed) to generate the whole
set of states. Without some additional physical axiom or some construction generating enough idealized "extremal" pure states, there
seems to be no compelling reason why there should exist any actually
preparable pure states at all, except as previously defined in the context
of an embedding.

Two positive results along these lines can be given, the first
obvious and the second very nearly so: mixed states of a QM embedding
are physically mixed states, and if there exists a set of physically pure
states that generates all the physically mixed states as convex combinations, then there exists a QM embedding in which all physically mixed
states are mixed states of the embedding.$^{(*)}$ It is the attempt to establish convincingly the "if" part of this last proposition that runs into
trouble. Fortunately, we seem to be able to do quantum theory quite
well without such a result.

We can go some distance, though, toward economizing in our
quantum theoretical model building, at least in principle, by a dimensionality argument: clearly there exists a minimum dimension that a
Hilbert space can possess and still be capable of supporting a QM
embedding of a given universe. Presumably this minimum dimension
would in some sense optimize both the number of mixed states and the
number of pure states expressible in terms of other pure states, and
if the requirement of minimal dimensionality did not uniquely determine
an embedding, we would be free then to apply additional criteria in
narrowing our choices, such as simplicity, elegance, ease of computation, etc. We emphasize the "in principle" caveat of these remarks,
since there seems to be no way of determining such a minimum
dimension, and even if there were, the more qualitative criteria
mentioned above might well override considerations of dimension.

---

(*) Cf. reference 2 for proof.

In practice we take the approach that, since the Hilbert space language can handle any physical universe expressed in terms of questions and states, the main task of physical theory is to find simple physical models that both incorporate all observed phenomena to date and make predictions for further research while using only what is judged in some sense to be a "most economical" class of physical systems, states and observables. We will not be overly bothered by the possibility that the Hilbert space mathematical language of QM embeddings may introduce idealized questions and states, since this seems to be a feature present in all known injections of empirical situations into logical or mathematical structures, and seems a small price to pay for the geometric clarity and computational simplicity of Hilbert space. Further questions concerning the dimensionality of the Hilbert space underlying an embedding will be dealt with as necessitated by the requirements of particular physical models, and to avoid unnecessarily fussy distinctions (until they are really needed) , we will simply assume, unless specifically stated otherwise, that any QM embeddings used have been chosen with maximum economy. This last assumption will manifest itself in our acceptance of the standard practice of treating all projections and nonnegative trace 1 operators in a QM embedding (modulo superselection rules) as physical questions and states, respectively, at least in some "potential" or "virtual" sense.

Returning now to a general analysis of conjunction or "interaction" of systems, for any embedding under construction let $\mathcal{S}$ be the projection from the tensor algebra $H^{\otimes}$ of the underlying Hilbert space onto the symmetric tensor algebra $H^{\odot}$, and let $\mathcal{A}$ be the corresponding projection onto the antisymmetric algebra $H^{\wedge}$. (For a good review of these notions and others we will be using, see reference 5). Several easily checked features of the embedding constructed in T9 above should now be noted:

a) The underlying Hilbert space $H^{\otimes}$ of the embedding is infinite dimensional, with $H^{\odot}$ of infinite dimension, and $H^{\wedge}$ of dimension $2^n - 1$ , $n = \dim (H) < \infty$ , and of infinite dimension otherwise. We put aside the question of the order of infinity involved in the dimensionality of $H^\otimes$ when H is nonseparable, since, as we will see, we do not need this case, and simply note that if H is separable, so is $H^\otimes$.

b)  If $H_n^\otimes$, $n = 0, 1, 2, \ldots$, is the family of nth-order tensor subalgebras of $H^\otimes$, with $\mathcal{P}_n$ the corresponding pairwise orthogonal family of projections on $H^\otimes$, and if $H_n^\odot = \mathcal{S} H_n^\otimes$, $\mathcal{S}_n = \mathcal{P}_n \mathcal{S} \mathcal{P}_n$, $H_n^\wedge = \mathcal{A} H_n^\otimes$, $\mathcal{A}_n = \mathcal{P}_n \mathcal{A} \mathcal{P}_n$ we have

$$\mathcal{A} \mathcal{S} = \mathcal{S} \mathcal{A} = \mathcal{P}_0 \oplus \mathcal{P}_1$$

$$\left. \begin{array}{rcl} \mathcal{S} \mathcal{P}_n = \mathcal{P}_n \mathcal{S} &=& \mathcal{S}_n \\ \mathcal{A} \mathcal{P}_n = \mathcal{P}_n \mathcal{A} &=& \mathcal{A}_n \end{array} \right\} \quad n \geq 0$$

$$\mathcal{S}_0 = \mathcal{A}_0 = \mathcal{P}_0$$

$$\mathcal{S}_1 = \mathcal{A}_1 = \mathcal{P}_1$$

$$\mathcal{S}_n \mathcal{A}_m = \mathcal{A}_m \mathcal{S}_n = 0, \quad n \geq 2 \text{ or } m \geq 2.$$

c)  Obviously $\underline{T9}$ has a great deal of redundancy in its representation of states:  besides the fundamental representatives $\varphi_S$ and $\eta_S$, guaranteeing that the embedding is $QM^2$, we have simply extended the projections $\pi_R$, representing questions, to the larger context of $H^\otimes$ in such a way that these projections have the proper trace on the products $\varphi_T \otimes \varphi_V \otimes \varphi_W \otimes \ldots$ in all cases where $T * V * W * \ldots \neq 0$. In terms of the operators $\eta_T$, $\eta_V$, etc. we have $\varphi_T \otimes \varphi_V \otimes \ldots = (\eta_T \otimes \eta_V \eta \ldots)(\eta_T^* \otimes \eta_V^* \otimes \ldots)$.

d)  By reason of the symmetry of the interaction product $*$ and the method of constructing the extended projections $\pi_R$, $R \in \mathfrak{Q}_U$, in $\underline{T9}$ (in such a way that each product space $H_T \otimes H_V \otimes \ldots$ is invariant under $\pi_R$), it is easily checked that, for any unit norm Schmidt operator of the form $\alpha_S = \sum\limits_{(T, V, \ldots)} C_{T, V, \ldots} \, \eta_T \otimes \eta_V \otimes \ldots$, where the summation is over ordered sets of states $(T, V, \ldots)$ such that $T * V * \ldots = S \neq 0$ for some fixed state S, we have trace $(\pi_R \alpha_S \alpha_S^*) = $ trace $(\pi_R \varphi_S)$, $\forall R \in \mathfrak{Q}_U$. In particular, for the normalized symmetric operators $\sqrt{n!} \, \mathcal{S} \, (\eta_T \otimes \eta_V \otimes \ldots) = \frac{1}{\sqrt{n!}} \sum\limits_{\sigma} \eta_{\sigma(T)} \otimes \eta_{\sigma(V)} \otimes \ldots$

or the antisymmetric operators of the form $\sqrt{n!}\ \mathcal{A}\ (\eta_T \otimes \eta_V \otimes \ldots) = \dfrac{1}{\sqrt{n!}}\ \sum\limits_{\sigma}\ \text{sgn}\ (\sigma)\ \eta_{\sigma(T)} \otimes \eta_{\sigma(V)} \otimes \ldots$ (summation over all permu-

tations $\sigma$ of order n) , or for properly normalized operators of arbitrary mixed symmetry, antisymmetric in some states, symmetric in others, the projections $\pi_R$ representing questions in $\underline{T9}$ will satisfy the appropriate trace formula.  Thus $\underline{T9}$ guarantees the mathematical existence of second quantized Hilbert space representations of the physical universe that include all the various symmetries encountered in modern theories of particles and fields, and we are free then, within this universal mathematical language, to choose further conventions, definitions, and physical assumptions for determining specific models for particular physical situations.

Modern physics, of course, has generally attempted to assign a definite symmetry to the representation of each composite system or state on the basis of transformation group representation theory, and all representatives of a given situation T * V * W * ... having a symmetry different from the one assigned are excluded as unphysical (e.g., systems whose composite states are assigned totally antisymmetric representatives are called fermions, while those with totally symmetric representations are called bosons, and the spin labels of group representation theory then assign half-add integer spins to fermion states and integer spins to boson states).  Most recently, supersymmetry theories have given signs of attempting to make use of previously excluded representations (to each boson there is a corresponding fermion, etc.).  The mathematical language developed here will, of course, be able to handle all such developments, since, by reason of the degree of orthogonality present in the construction used, $\underline{T9}$ guarantees the existence of embeddings in which the trace formula for each projection $\pi_R$ , $R \in \mathfrak{Q}_U$ , can be extended <u>independently</u> to different symmetry representations of the same composite state, even though the particular extension explicitly constructed there happens to assign the same trace value for each $\pi_R$ on all the symmetry assignments for a given composite state.

e) Lastly, we note that the particular assignment in $\underline{T9}$ of projections on $H^{\otimes}$ to questions gives the following trace values on the operators representing "unphysical" situation such as $T * V * W * \ldots = 0$ :

$$\forall\, R \in \mathfrak{D}_U \,, \quad \text{trace} \left( \pi_R\, \varphi_T \otimes \varphi_V \otimes \varphi_W \otimes \ldots \right) = \begin{cases} 0 \,, & R \in \mathfrak{m} \\ 1 \,, & R \notin \mathfrak{m} \,, \end{cases}$$

where, of course, $\mathfrak{m}$ is the base of $\mathfrak{D}_U$ used for the original embedding with respect to H. It is easily seen, however, that the construction used in $\underline{T9}$ guarantees that we can extend the projections $\pi_R$ to $H^{\otimes}$ in such a way that the trace formula above takes on any values required by some physical model in which we might choose to give an interpretation to the operators $\varphi_T \otimes \varphi_V \otimes \ldots$ corresponding to situations $T * V * \ldots = 0$, i.e., these operators are completely free elements in our language as constructed so far.

## III. A Physical Model Incorporating Relativity

At this point we need several further definitions and physical assumptions. We will progress in relatively small steps, so that the precise effect each axiom has on the previously existing structure (and possible consequences of modifying a specific assumption) can be considered more or less in isolation.

$\underline{D11.}$ Let $\overline{L} \subset \mathcal{P}(Q)$ be the class of all $L \in \mathcal{P}(Q)$ such that
$$\forall\, R \in \mathcal{P}(Q) \,, \quad P\,(* [R \cup L]) = P\,(R).$$
Call each $L \in \overline{L}$ an <u>observing station</u> or simply a <u>lab</u>. $\square$

Obviously, for each lab L, in the functional representation of questions, we have $\widetilde{L} = \widetilde{1}$, and in the functional representation of systems, $\widehat{L} = \widehat{1}$. Trivially, since $\emptyset \in \mathcal{P}(Q)$ has the required properties, $\overline{L}_U$ is not empty, but of course we will be thinking in terms of a universe in which there exists a great variety of labs, each to be conceived of as a sort of "window on the universe" or " way of looking at the universe." Concretely, one can think of each $L \in \overline{L}$ as a collection of lists of instructions for constructing a completely computerized "black box" of instrumentation, preprogrammed to carry out, on command, the instructions necessary for asking, and recording the answer of, all possible questions on the universe. Type out the proper access code on the console built into such a lab and it sets about performing and recording the result of the corresponding question. This notion of lab generalizes and replaces the automated observer-clock-rod frame discussed, for example, in our reference 6 , p. 17, ff., since each lab can make all measurements, and is not limited to space and time determinations only.

Once a notion of intercommunication is injected into our interpretation, the various labs may even be thought of as differing widely in particularities (think in terms of such as yet undefined notions as position, velocity, temperature, energy content, mode of construction, etc.), as long as each is required to be able to ask all the questions any other can ask, if necessary via intercommunication so that each has access to any data another might acquire by reason of some special status or circumstance. [A telescopic observation of the surface texture of Mars made from Mount Palomar must agree with results of an on-site inspection by a spacecraft landing on Mars if the two experiments are to be said to be asking the same questions, but some questions may only be able to be asked by the spacecraft, and then Palomar will have to be content with making the questions its own via telemetry].

Another way of expressing the main thrust of $\underline{D11}$ is to say that the functional representations $\widetilde{\mathfrak{D}}_U$ and $\hat{S}_U$ are defined by the sum total of all the probabilistic information acquired in all the labs, shared freely by all. Any differences between labs gets washed out in the full functional representations $\widetilde{\mathfrak{D}}_U$ and $\hat{s}_U$.


$\underline{D12}$. A chronology on $\overline{L}$ will mean a mapping $C$ that assigns to each $L \in \overline{L}$ a family $C(L) = \{ C_r (L) \subset \underset{\sim}{P}(Q) : C_r (L) \neq \emptyset , r \in \mathcal{J}_C \}$, where the index set $\mathcal{J}_C$ for the families is some subset of the real numbers $\mathbb{R}$, such that, $\forall L \in \overline{L}$ ,

$$1) \quad \bigcup_{r \in \mathcal{J}_C} C_r (L) = \mathfrak{m} \text{ (a base for } \mathfrak{D} \text{ - - cf. D5);}$$

2) $\forall\ r, r' \in \mathcal{J}_{C}\ ,\ \ r \le r' \ \Rightarrow \ \mathcal{C}_{r}\,(L) \subset \mathcal{C}_{r'}\,(L)$ ;

3) $\forall\ L^{1} \in \overline{L}\ ,\ \ \forall\ r \in \mathcal{J}_{C}\ ,$

$$\exists\ r',\ r'' \in \mathcal{J}_{C}\ :\ \mathcal{C}_{r'}(L) \subset \mathcal{C}_{r}\,(L^{1}) \subset \mathcal{C}_{r''}\,(L).$$

Assuming that some chronology $\mathcal{C}$ has been fixed, we will usually simplify our notation by writing $L_{r}$ for $\mathcal{C}_{r}\,(L)$ and $\mathcal{J}$ for $\mathcal{J}_{C}$. Call $L_{r}$ lab L at proper time r and $\mathcal{J}$ the proper time parameter of the chronology; for $N \subset \mathcal{P}_{\sim}(Q)$ , say that N is in the past of $L_{r}$ if $N \subset L_{r}$. For L, $L^{1} \in \overline{L}$ we allow various obviously synonymous phrases for the relation $L^{1}_{r'} \subset L_{r}$ , e.g., $L_{r}$ is in the future of $L^{1}_{r'}$ ; $L_{r}$ has past access to or past communication with $L^{1}_{r'}$ ; $L^{1}_{r'}$ has future access to or future communication with $L_{r}$ . For $\mathcal{g} \subset \mathcal{J}$ , let $\overline{L}_{\mathcal{g}} = \{ L_{r}\ :\ L \in \overline{L}\ ,\ r \in \mathcal{g}\ \}$, and define the proper time coordinate pair in L of $L^{1}_{r}$ , in symbols $p_{L}\,(L^{1}_{r}) = (\ p^{1}_{L}\,(L^{1}_{r})\ ,\ p^{2}_{L}\,(L^{1}_{r})\ ) \in \mathcal{R} \times \mathcal{R}$, by: $\ \ \forall\ L \in \overline{L}\ ,\ \ L^{1}_{r} \in \overline{L}_{\mathcal{g}}\ ,$

$$p^{1}_{L}\,(L^{1}_{r}) = \sup\ \{\ r' \in \mathcal{J}\ :\ L_{r'} \subset L^{1}_{r}\ \}$$

$$p^{2}_{L}\,(L^{1}_{r}) = \inf\ \{\ r' \in \mathcal{J}\ :\ L^{1}_{r} \subset L_{r'}\ \}$$

[ Note that $p_{L}\,(L^{1}_{r})$ is not necessarily in $\mathcal{J} \times \mathcal{J}$. ]   Define the functions $s_{L}\ :\ \overline{L}_{\mathcal{J}} \to \mathcal{R}$ and $t_{L}\ :\ \overline{L}_{\mathcal{J}} \to \mathcal{R}$ by

$$s_{L}\,(L^{1}_{r}) = \frac{1}{2}\ [\,p^{2}_{L}\,(L^{1}_{r}) - p^{1}_{L}\,(L^{1}_{r})\,]$$

$$t_{L}\,(L^{1}_{r}) = p^{1}_{L}\,(L^{1}_{r}) + s_{L}\,(L^{1}_{r})$$

$$= \frac{1}{2}\ [\,p^{2}_{L}\,(L^{1}_{r}) + p^{1}_{L}\,(L^{1}_{r})\,].$$

Let $n^{r}_{L}\ :\ \overline{L} \to Z^{+}_{\infty}$ (extended nonnegative integers ), $L \in \overline{L}$ , $r \in \mathcal{R}$ , be defined by the rule that the range value $n^{r}_{L}\,(L^{1})$ is the cardinality of $\{\,r' \in \mathcal{J}\ :\ p^{2}_{L}\,(L^{1}_{r'}) = r\ \}$ when this set is finite, and $n^{r}_{L}\,(L^{1}) = \infty$ otherwise.   Call $p^{1}_{L}$ , $p^{2}_{L}$ the first and second proper time functions in L, $s_{L}$ the radial space coordinate in L ( or simply the space coordinate in L when no confusion is likely), $t_{L}$ the time coordinate in L, and $n^{r}_{L}$ the frequency count in L at proper time r.   Define the spatial tube of radius r in L as the set

$$\Sigma^{r}_{L} = \{\,L^{1}_{r''} \in \overline{L}_{\mathcal{J}}\ :\ s_{L}\,(L^{1}_{r''}) = r\,\}\ ,$$

the <u>spatial sphere of radius</u> r <u>in L at time</u> r ′as

$$\Sigma^{r}_{L_r\prime} = \{\, L^1_{r\prime\prime} \in \overline{L}_{\mathcal{J}} \,:\, s_L(L^1_{r\prime\prime}) = r,\ t_L(L^1_{r\prime\prime}) = r\prime \,\},$$

and the <u>celestial   sphere in L at proper time</u> r (also, <u>light cone surface</u>
<u>in L at</u> r) as

$$\Lambda^{L}_{r} = \{\, L^1_{r\prime} \in \overline{L}_{\mathcal{J}} \,:\, p^2_L(L^1_{r\prime}) = r \,\}.$$

When we need to consider various chronologies $C$, $C\prime$, ... , we will
simply replace $L_r$ by $C_r(L)$ , L by $C(L)$ , etc., at appropriate points
in the above notation.      □


The definitions in <u>D12</u> are fairly self-explanatory, and in
choosing them we have obviously allowed ourselves to be guided by
standard relativity theory.  The acceptance of such guidance whenever
available will be adopted as a basic methodological principle without
much apology, since we view it as simply a means of building as much
geometry into our structure as possible, and the rather continuously
successful history of geometry in physical theory should allow such a
policy to stand on its own merits.

Our motivation, of course, is the need to introduce the notions
of time and communication into our formal structure.  To help our
imagination we can think of each computerized observing apparatus L
as being equipped with a time-keeping device : some operationally
definable means (to be specified further) for parameterizing by real
numbers the questions asked by L.  Each $L_r$ can then be thought of
as the past light cone in L at time r, consisting of all the probabilistic
information stored in L up to r.  Then <u>D12</u>, 1) , 2) and 3) simply spell
out the characteristics expected of such a parameterization, and the
remainder of <u>D12</u> specifies the notions needed for defining space time
in terms of proper time.

We have now:


<u>T10</u>.      For $L_r$ , $L^1_{r\prime}$ , $\in \overline{L}_{\mathcal{J}}$ , all of the conditions below are true if and
         only if any one of them is true:

1) $L_r \in \Lambda_r^{L^1}$,  and $L_r^1{}' \in \Lambda_r^{L}$

2) $p_L^2 (L_r^1{}') = r$ and $p_{L^1}^2 (L_r) = r'$

3) $p_L^1 (L_r^1{}') = r$ and $p_{L^1}^1 (L_r) = r'$

4) $p_L (L_r^1{}') = (r, r)$

5) $p_{L^1} (L_r) = (r', r')$

6) $s_L (L_r^1{}') = 0$ and $t_L (L_r^1{}') = r$

7) $s_{L^1} (L_r) = 0$ and $t_{L^1} (L_r) = r'$

8) $L_r = L_r^1{}'$ .  $\square$

Proof: That 4) through 8) are equivalent to one another follows immediately from the definitions of $p_L$ , $s_L$ and $t_L$ in <u>D12</u>. But, clearly, 4) and 5) together imply 3), and also 4) and 5) together imply 2), which in turn implies 1). Since 4) and 5) are equivalent conditions, either one alone is a sufficient condition for each of statements 1), 2), 3) . But condition 2) is simply a restatement of 1), so they are equivalent. Finally, since 2) implies both $L_r^1{}' \subset L_r$ and $L_r \subset L_r^1{}'$ , i.e., 2) $\Rightarrow$ 8) , and analogous reasoning shows that 3) $\Rightarrow$ 8), and 8) is equivalent to 4), the theorem is established.                     Q.E.D.


<u>T11</u>.    For $L_r$ , $L_r^1{}' \in \bar{L}_{\mathcal{J}}$ , $\Lambda_r^{L} = \Lambda_r^{L^1}{}'$ implies each, and therefore all, of the equivalent conditions 1) through 8) of <u>T10</u>.     $\square$

Proof: $\Lambda_r^{L} = \Lambda_r^{L^1}{}'$ obviously implies <u>T10</u>, 1), and so the theorem follows:    Q.E.D.


The situation we would like to investigate now is that in which the converse of <u>T11</u> holds .

$\underline{D13}$.      Say that a chronology $C$ is relativistic if, $\forall \; L_r$, $L_r^1$, $\in \overline{L}_{\mathcal{J}}$, $[L_r = L_r^1,] \Rightarrow [\Lambda_r^L = \Lambda_r^{L^1},]$, and, for two chronologies $C$, $C'$, define $C$ to be <u>less than</u> $C'$ (in symbols $C \leq C'$) if, $\forall \; L \in \overline{L}$,

     1)   $C(L) \subset C'(L)$,

     2)   $\forall \; r \in \mathcal{J}_C$, $\Lambda_r^{C(L)} \subset \Lambda_r^{C'(L)}$,

where, of course, $\Lambda_r^{C(L)}$ means the celestial sphere in L at r as determined by chronology $C$. Condition 1) makes use of the standard inclusion ordering of mappings (sets of pairs -- in this case, then, each family $C(L)$ is simply the set $\{(r, C_r(L)) : r \in \mathcal{J}_C\})$, and so implies $\mathcal{J}_C \subset \mathcal{J}_C$, for $C \leq C'$. Thus condition 2) makes sense, and requires that the light cone surface in lab L at time r as determined in chronology $C'$ should contain the light cone surface in L at r as determined in $C$ if $C \leq C'$. An equivalent way of stating 1) is:

$$\forall \; L \in \overline{L}, \; C(L) = C'(L) \mid \mathcal{J}_C .$$

A chronology $C$ will be said to be <u>maximal</u> if, for arbitrary chronologies $C'$, $C \leq C' \Rightarrow C = C'$, and $C$ will be said to be <u>maximal relativistic</u> if $C$ is relativistic and $C \leq C' \Rightarrow C = C'$ for arbitrary relativistic chronologies $C'$.     $\square$


$\underline{T12}$.      There exist maximal chronologies and maximal relativistic chronologies in any nonempty universe.     $\square$

Proof:    That chronologies exist, and in great abundance, can be seen from easily checked examples, e.g., trivially, let $\mathcal{J} = \{r\} \subset \mathcal{R}$ and, $\forall \; L \in \overline{L}$, set $L_r = \mathbb{m}$. Since the chronology so defined is relativistic, existence of relativistic chronologies is settled also. Less trivially, let $\mathcal{J} \subset \mathcal{R}$ be finite with $r_0 \in \mathcal{J}$ as least element, and for each $L \in \overline{L}$ let $L_{[r]}$ be a family of nonempty subsets of $\mathbb{m}$ such that

$$\mathbb{m} = \bigcup_{r \in \mathcal{J}} L_{[r]} \quad \text{and } r \neq r' \Rightarrow L_{[r]} \neq L_{[r']}, \quad \text{with } L_{[r_0]} = L_{[r_0]}^1, \; \forall \; L^1 \in \overline{L}. \quad \text{Set } L_r = \bigcup_{\substack{r' \in \mathcal{J} \\ r' \leq r}} L_{[r']}. \quad \text{Then } C_r(L) =$$

$L_r$ defines a chronology; if, in addition, $L_r^1 = L_r$, $\forall\, L,\, L^1 \in \bar{L}$, $\forall\, r \in \mathcal{J}$, the chronology is relativistic.

It is obvious that $\leq$ is a partial ordering on chronologies, and straightforward to check that, if $C^{\cup}(L) = \bigcup_i C^i(L)$ for some chain of chronologies $C^i$, then $r \to C_r^{\cup}(L)$, $r \in \mathcal{J}_{C^{\cup}} = \bigcup_i \mathcal{J}_{C^i}$, is a chronology, which we will call the <u>union chronology</u>. It is also easy to check that <u>D13</u>, 2) would be violated if, for some $C^j$ in a chain,

$$\Lambda_r^{C^i(L)} \not\subset \Lambda_r^{C^{\cup}(L)} .$$

Hence, $C^{\cup}$ is an upper bound for the chain $C^i$, and by Zorn's Lemma, there exist maximal chronologies. Similar reasoning shows that, for a chain $C^i$ of relativistic chronologies, if the union chronology is not relativistic, then the relativistic condition is violated for some $C^j$ in the chain, contrary to hypothesis, and so, again by Zorn's Lemma, there exist maximal relativistic chronologies.            Q. E. D.


To summarize, from the point of view of this paper a chronology simply provides for each lab L a means of dividing a base $\mathfrak{m}$ for the questions into subsets $L_r$ ordered by inclusion and labeled by a real parameter $r \in \mathcal{J}$. If $\Omega$ is a parameter set labeling distinct labs, then in our model so far the sets $L_r^{\alpha}$, $\alpha \in \Omega$, $r \in \mathcal{J}$, are the analogs of lightcones in standard relativity theory, and since lightcones are in one-to-one correspondence with events regarded as apexes of lightcones, we can interpret the objects $L_r^{\alpha}$ as the analogs in our model of spacetime points. Thus the set $\bar{L}_{\mathcal{J}}$ becomes the raw material from which we must construct spacetime. To help in this construction we have one relation on $\bar{L}_{\mathcal{J}}$, inclusion ($L_r^{\alpha} \subset L_r^{\beta}$ interpreted as $L_r^{\alpha}$ in the past of $L_r^{\beta}$), and the fact that in the context of any given lab L inclusion is a total ordering and is faithfully reflected by the ordering of the parameter values $r \in \mathcal{J}$. For the whole set $\bar{L}_{\mathcal{J}}$, of course, inclusion is only a partial ordering, but, as worked out in detail in reference    7    , this is already enough structure to build topology from, and we can hope that further rea-

sonable physical requirements might elucidate the origin of the macroscopically observed topological properties of spacetime. (*)

We investigate now how chronologies look in the context of QM
embeddings. Since each $L_r \in \bar{L}_{\mathcal{J}}$ is a set of nonnull questions in $\mathcal{Q}_U$,
we essentially have to consider the image sets $\pi(L_r)$ and the corres-
ponding sets of state images under the mapping $\varphi$, where $\{\pi, \varphi\}$ is
some QM embedding, along with any useful mathematical objects that
can be constructed using these sets.

<u>D14</u>.        Let $\Omega$ be a set of labels $\alpha$, $\beta$, $\ldots$, $\omega$ for labs, which we will
write as superscripts (i.e., $L^\alpha$, $L^\beta$, $\ldots \in \bar{L}$). Throughout the remainder of this paper we assume that some chronology has been fixed,
and all symbols such as $L_r^\alpha$ and $\mathcal{J}$ refer to that chronology. For
$\{\pi, \varphi\}$ a QM embedding of the universe with respect to a Hilbert space
H and $L_r^\alpha$ an element of $\bar{L}_{\mathcal{J}}$, let $T_r^\alpha$ be the projection on H corresponding to the subspace union of the ranges of the projections in the set
$\pi(L_r^\alpha)$, or, in abbreviated notation,

$$T_r^\alpha = \bigvee_{R \in L_r^\alpha} \pi R,$$

and define the projection $\tau_r^\alpha$ by

$$\tau_r^\alpha = T_r^\alpha - \bigvee_{\substack{r' \in \mathcal{J} \\ r' < r}} T_{r'}^\alpha.$$

Finally, let $\mathcal{B}$ be the Borel sets of the real line, and define

---

(*)   Our motivation for following the lead of standard relativity theory
wherever possible has already been discussed in connection with <u>D12</u>,
but our particular choice of definition for the term "relativistic" in
<u>D13</u> warrants some comment. Basically it is rooted in an idea due to
Komar[8]: relativity can in large part be imposed by requiring that
observers at the same event see the same celestial sphere.

the family of projections $\tau^{\alpha}_{B}$ , $B \in \mathcal{B}$ , by

$$\tau^{\alpha}_{B} = \bigvee_{r \in B} \tau^{\alpha}_{r} \; .$$

If the family $\tau^{\alpha}_{B}$ defines a projection-valued measure we call the corresponding self-adjoint operator $\tau^{\alpha}$ the <u>proper time operator</u> for lab $L^{\alpha}$ in the embedding $\{ \pi, \varphi \}$ .  □

We have as a basic result:

T13.    For any QM embedding of a universe with chronology the proper time operator $\tau^{\alpha}$ exists for each lab $L^{\alpha} \in \overline{L}_{\mathcal{J}}$ .  □

Proof:    We must establish that the projections $\tau^{\alpha}_{B}$ , $B \in \mathcal{B}$ , of D14 form a projection-valued measure, i.e., that

1)  $\tau^{\alpha}_{\varnothing} = 0$  and  $\tau^{\alpha}_{R} = I$ ;

2)  $B \cap C = \varnothing \;\Rightarrow\; \tau^{\alpha}_{B} \perp \tau^{\alpha}_{C}$ ;

3)  $B_{i} \cap B_{j} = \varnothing$ for $i \neq j \;\Rightarrow\;$

$$\tau^{\alpha}_{\underset{i}{\cup} B_{i}} = \sum_{i}{}^{\oplus} \tau^{\alpha}_{B_{i}} \; .$$

Condition 1) is obvious from D14 and the definition of a chronology (D12) since every element of a base $\mathfrak{m}$ , including $\tilde{1}$ , must appear in some $L^{\alpha}_{r}$ . Also, from the definition of $\tau^{\alpha}_{r}$ in D14 , it is immediate that $r \neq r'$ implies $\tau^{\alpha}_{r} \perp \tau^{\alpha}_{r'}$ . Hence 2) and 3) follow easily.    Q.E.D.

Of course, unless the chronology is in some sense natural or "realistic," the proper time operator, while existing, will not have much significance. For example if some $L^{\alpha}_{r}$ contains the identity question $\tilde{1}$ then no $r'$ greater than r can be in the spectrum of $\tau^{\alpha}$ —— hence inclusion of $\tilde{1}$ in $L^{\alpha}_{r}$ should be put off until "the end of time." Leave aside the technical problems with this term for the moment,

since our remaining physical assumptions will resolve them and further
specify what a realistic chronology should mean.  At any rate the existence of the proper time operators can help us in motivating and formulating our final model.  For example, assuming that $\mathcal{J}$ is the spectrum of each operator $\tau^\alpha$ and that a conjugate self-adjoint operator
$\mu^\alpha$ exists for each $\tau^\alpha$ (to be thought of as the generator of motion along
the spectrum of $\tau^\alpha$) , four general possibilities immediately suggest
themselves:

1)   Continuous spectrum for $\tau^\alpha$ and $\mu^\alpha$ .

2)   Continuous spectrum for $\tau^\alpha$ ; discrete spectrum
     for $\mu^\alpha$.

3)   Discrete spectrum for $\tau^\alpha$; continuous spectrum
     for $\mu^\alpha$ .

4)   Discrete spectrum for $\tau^\alpha$ and $\mu^\alpha$.

The first three situations all obviously require infinite-dimensional
Hilbert space, while 4), with any reasonable definition of conjugacy,
can only hold if we have a finite dimensional space for our basic QM
embedding.

We choose to limit the possible models for the physical universe to options 3) and 4) above, and we will in fact use an integer
model for the proper time parameter set $\mathcal{J}$ , while recognizing that
this breaks rather radically with the usual continuum methods for in-
corporating time into quantum theory.  This very feature, however, can
be viewed as part of the appeal of such a model, since it should offer
a greater possibility of healing the fundamental ailments of the more
usual methods, and it also seems to remain more faithful to the premises of a fully operational approach to physical theory:  actual operational methods for determining time always seem to come down to a
simple matter of counting.  Hence we propose to axiomatize the
dictum "God knows how to count, but He doesn't do calculus" (attributed
in this form to Feynman).  We recognize that we may well have to do
calculus, if only as the most convenient approximation available for
making the transition from the extremely large numbers of microscopic
physics to the smooth appearances of the macroscopic world.

In choosing the remaining assumptions needed to determine a specific model, we will be guided by two basic principles:

A)　we want to extend the well understood and tested features of macroscopic spacetime as far as possible into the realm of the submicroscopic;

B)　we recognize that all available evidence indicates that such extension must fail at some point and yield to general quantum principles.

Hence we think of our labs $L^\alpha$ as being microminiaturized to the ultimate degree, so much so in fact that each new lab instant $L^\alpha_r$ can bring with it only the possibility of asking exactly one new question which in turn defines one new state, to be interpreted as the state of the universe determined locally along the world line of lab $L^\alpha$ at proper time instant r. In the most simple minded QM embedding of such a model, each new state and question will be identified with one another as a one-dimensional projection $\tau^\alpha_r$ representing the corresponding lab instant. To further help our imagination we can take the viewpoint of modern particle physics, and regard each ultimately miniaturized lab $L^\alpha$ as a spatially structureless entity (operationally definable as exhibiting no internal structure under, say, arbitrary scattering experiments) pumped in such a way that it emits detectable pulses (beats, beeps) $\tau^\alpha_r$ which are, by definition, taken to be regular, and which we can refer to as the chronons or instantons of $L^\alpha$. Thus each $L^\alpha$ is to be regarded both as an ultimate operationally definable world line and as the ultimate time and frequency standard (standard clock) along the world line. All macroscopic questions, states, observing apparatus (and observers) will be built up from large numbers of the fundamental states and questions constituting ultimate world lines and combinations (interactions) of which these states and questions are capable.

As preliminaries to the formal statement of our third axiom we have:

<u>D15.</u> Define a chronology to be <u>integer</u> if the proper time parameter set

$\mathcal{J}$ is a subset of Z, and say that a chronology is <u>simple</u> if there exists a coherent universal $QM^2$ embedding (with the base $\mathfrak{m}$ for the embedding also the base for the chronology) such that the proper time operators $\tau^\alpha$ represent observables and share a common simple spectrum $\mathcal{J}$.   $\square$

The integers of course bring with them their characteristic ordering, and this removes a troublesome possibility inherent in the general definition of chronology (<u>D12</u>)   The relevant facts are stated as:

<u>T14.</u>        For an integer chronology the proper time coordinate functions $p_\alpha^1$, $p_\alpha^2$, corresponding to any lab $L^\alpha$, have ranges in $\mathcal{J}$, while the functions $s_\alpha$ and $t_\alpha$ have integer or half-integer values. Moreover, for $r \notin \mathcal{J}$, the frequency count functions $n_\alpha^r$, $\alpha \in \Omega$, are identically zero.        $\square$

Proof:       Immediate from the definitions of functions in <u>D12</u>.   Q.E.D.

We have further:

<u>T15.</u>        For a simple integer chronology there exists a coherent universal $QM^2$ embedding of the universe with respect to separable Hilbert space.       $\square$

Proof:      By <u>D15</u> there exists an embedding of the required type in which each operator $\tau^\alpha$ has simple integer spectrum, but for self-adjoint operators this can only occur in a separable Hilbert space, since the eigenvectors of $\tau^\alpha$ must span the space.       Q.E.D.

In the terminology developed above we can state our third physical axiom as

<u>U3.</u>        There exists a maximal relativistic simple integer chronology for the universe.       $\square$

For our final axiom we need some additional terminology:

<u>D16</u>.       Assume from now on that a chronology of the type postulated by <u>U3</u> has been fixed, together with an embedding as specified in <u>T15</u>. Say that a projection $\pi$ on a Hilbert space H is <u>orthocyclic</u> with respect to an operator A if the ranges of the operators $A^n \pi$, $n = 0, 1, 2, \ldots$, span H and

$$A^n \pi \neq A^m \pi \Rightarrow \pi A^{*m} A^n \pi = 0.$$

[This combines the notion of a cyclic projection with that of a wandering subspace (Cf. Ref. 9, pg. 77).]       □

As our last axiom we have:

<u>U4</u>.       For all proper time operators $\tau^\alpha$, $\alpha \in \Omega$, the projection $\varphi_1$ representing the universal state is orthocyclic with respect to the unitary operators $U_\alpha = e^{i k \tau^\alpha}$, where k is a universal constant.       □

An immediate result is

<u>T16</u>.       The Hilbert space of an embedding satisfying <u>U4</u> is finite dimensional.       □

Proof:     Writing $\varphi_1 = |1\rangle \langle 1|$, the definition of orthocyclicity requires that the sequence $U_\alpha^n |1\rangle$, $n = 0, 1, 2, \ldots$, contain an orthonormal basis for H. If H were infinite-dimensional then $U_\alpha$ would be a simple unilateral shift on the resulting basis, and it is well known that such an operator can have no eigenspaces (Cf. Ref. 9 for relevant definitions and proofs), contradicting our assumption of discrete simple spectrum for $\tau^\alpha$. Hence H cannot be infinite-dimensional. To see that the required phenomenon can occur in a finite-dimensional space, we can require that each $U_\alpha$ be a simple "round robin shift," i.e., a unitary operator with orthocyclic vector $|1\rangle$ such that, for N the dimension of H, $U_\alpha^N |1\rangle = |1\rangle$. The self-adjoint generator $\tau^\alpha$ for any such $U_\alpha$ will of course have the required discrete (finite) spectrum. Q.E.D.

The finiteness imposed by $\underline{U4}$ on the proper time parameter $\mathcal{T}$ resolves the previously mentioned difficulty with "the end of time," and we also have a precise meaning for the notion of conjugacy: a self-adjoint operator $\mu^{\alpha}$ is said to be <u>conjugate</u> to a proper time operator $\tau^{\alpha}$ if the unitary operator $V_{\alpha} = e^{-ik\mu^{\alpha}}$ is a simple "round robin" shift on a basis $|\tau^{\alpha}_{r}\rangle$ of eigenvectors of $\tau^{\alpha}$ ordered by the eigenvalues $r \in \mathcal{T}$. (This defines conjugacy up to a phase or gauge.)

It is the type of model universe specified by $\underline{U3}$ and $\underline{U4}$ that we propose for further investigation.

## IV. <u>Conclusion and Prospectives</u>

At first thought the restriction to a finite-dimensional universal Hilbert space might seem unacceptable, but on reflection one realizes that all the standard features of infinite-dimensional quantum theory are either present in the proposed model or are approximated as closely as desired for large enough dimension, and in fact we will require as a sort of correspondence principle that all the tested results of infinite-dimensional theory be reproduced in our model for suitably large dimension. If one wishes to insist on continuous time and countably infinite-dimensional Hilbert space, then weakening $\underline{U3}$ by allowing $\mathcal{T} = \Re$ (and slightly modifying the notion of chronology to handle the technical difficulty with the end of time), while maintaining $\underline{U4}$ intact, will provide such a model. From the point of view of the present paper, however, any initial assumption of continuous time (except as a quasi-classical approximation —— what Finkelstein has called the cq model: classical time, quantum everything else) would seem to be unnecessary and hence a violation of the principle of economy. If we find later that experiment or theory compels us to accept continuous time, our model is sufficiently flexible to incorporate it without strain.

In short, nothing seems to be lost in a finite discrete time model and much seems to be gained. One strongly attractive advantage is the extremely simple but richly elegant mathematical structure of conjugate pairs of unitary shifts on finite-dimentional spaces. More physically, a time and observer invariant vacuum state ($|1\rangle$) is built into our model, and vacuum expectation values form the basis for all probability calculations, with finite results guaranteed _a priori_ by the finite-dimensionality of the universal Hilbert space. It could be expected that Wheeler's "gates of time"[10] would appear naturally in our model —— the "big bang" and the "big crunch" being identified as a single ortho-cyclic projection $\tau_o^\alpha$ (an instanton) for the shift $V_\alpha$, approached from the two distinct directions of time along a worldline. A goal of Eddington, Dirac and others would also seem to find a natural setting in our model: the fundamental dimensionless constants of nature should be able to be computed in terms of the interplay between the dimensions of the universal Hilbert space and the various observed substructures. At minimum we might expect that some new algebraic classificational schemes for elementary particles and fields may become apparent from the restrictions an overall maximum dimension would put on allowable substructures.

To see how the machinery of special and general relativity might come out of our model, we need only require that all observers sharing a comon lab-instant come to agreement by convention on some uniform proper time coordinatization of the events in their common past. If (with a good dose of relativistic foresight, of course) they choose a coordinatization such that the product $p_\alpha^1 p_\alpha^2 = t_\alpha^2 - s_\alpha^2$ is the same for all the labs, then a two-dimensional Lorentz-invariant interval becomes available, and this interval is nonnegative on all the events in the common past of the observers. In terms of the Hilbert space structure of our model, if $H_+$ represents the subspace generated by the projections representing questions and states in the common past of the observers, then $t_\alpha^2 - s_\alpha^2$ should be represented on any of the world lines by a nonnegative operator $\sigma^2$ on $H_+$, which of course defines a nonnegative square root operator $\sigma$. Using the precise analog of Dirac's original derivation of the Dirac equation from the Klein-Gordon equation, we find that the existence of eigenvectors for $\sigma$ implies a distinguished complex quaternionic algebraic substructure on $H_+$ at each instant. Since the complex quaternions are isomorphic to the Pauli algebra, it can be hoped that a very natural derivation of the observed properties of space will result from the proposed model.    Perhaps equally significantly, once a complex 4-dimensional Hilbert space structure appears at each instant, it is natural to investigate twistors[11] (essentially one-dimensional projections on complex 4-space) as likely candidates for the fundamental projections $\tau_r^\alpha$.  This should allow a full incorporation of special relativity within our model.  We could then expect that a similar incorporation of general relativity into quantum theory should arise from the global analytic manifold structure of the full complex projective space of states (an essentially nonlinear structure) within the linear structure of the universal Hilbert space.  We hope to be reporting on more specific developments along these lines in the near future.


ACKNOWLEDGMENT

Our sincerest gratitude goes to Carl H. Brans for many helpful discussions.


REFERENCES

1.    A. R. Marlow, "An Extended Quantum Mechanical Embedding Theorem," this volume.

2.    A. R. Marlow, "Quantum Theory and Hilbert Space," J. Math. Phys. 19(9), 1841-1846 (1978).

3.    G. W. Mackey, "Mathematical Foundations of Quantum Mechanics," W. A. Benjamin, 1963.

4.    A. R. Marlow, "Orthomodular Structures and Physical Theory," in "Mathematical Foundations of Quantum Theory" (A. R. Marlow, Ed.), Academic Press, 1978.

5.    Robert Hermann, "Vector Bundles in Mathematical Physics," Vol. I, Ch. IX, W. A. Benjamin, 1970.

6.    E. F. Taylor, J. A. Wheeler, "Spacetime Physics," W. H. Freeman, 1966.

7.   A. R. Marlow, "Empirical Topology," to be published, <u>Int. J. Theor. Phys</u> (1980).

8.   Arthur Komar, "Foundations of Special Relativity and the Shape of the Big Dipper," <u>Am. J. Phys.</u> 33 (12), 1024 - 1027 (1965).

9.   P. R. Halmos, "A Hilbert Space Problem Book," p. 73 Van Nostrand, 1967.

10.  J. A. Wheeler, "Pregeometry: Motivations and Prospects," this volume.

11.  Roger Penrose, "Twistor Theory, Its Aims and Achievements," in "Quantum Gravity, an Oxford Symposium," (C. J. Isham, R. Penrose and D. W. Sciama, eds.) Oxford Univ. Press, 1975.

# AN EXTENDED QUANTUM MECHANICAL EMBEDDING THEOREM

A. R. Marlow

Department of Physics

Loyola University

New Orleans, LA 70118

1.    <u>Introduction.</u>

At the suggestion of R. J. Greechie,[1] the results of an earlier paper[2] (hereafter referred to as I) have been examined for possible extensions to more general situations. The earlier results established that arbitrary orthoposets $Q$ could be embedded in a measure-preserving way into both Boolean algebras and Hilbert orthologics if the embedding is required to preserve ordering only on a maximal orthogonal-exclusive subset $S \subset Q$. (Since this result implies that the ordering is also preserved on $Q - S = S'$, the set-theoretical complement of $S$, which happens in this case also to be the set of orthocomplements of elements in $S$, there is more preservation of order than might at first meet the eye, but the fact remains that complete order-preservation cannot be required.)

The result of the reexamination is that all constructions, proofs and theorems of I remain valid if what we define below as bounded dual posets everywhere replace orthoposets as domains of embeddings, and maximal orthogonal-exclusive subsets are generalized to maximal



Copyright © 1980 by Academic Press, Inc.
All rights of reproduction in any form reserved.
ISBN 0-12-473260-7

unitary dual-exclusive subsets.  We can then summarize everything
in an extended embedding theorem valid for an arbitrary dual poset
and its set of probability measures.

We believe these extended results are of more than purely
mathematical interest, and in the last section we discuss some
possible implications for the foundations of quantum theory.  Since
the present paper is not intended to be read independently of I, we
refer the reader to that previous paper for all notation, definitions
and results not explicitly  detailed below.

2.   The Extended Results

The additional definitions not already given in I are collected
together here as

D.   A pair $\{Q,'\}$ consisting of a poset $Q$ and a mapping $q \to q'$ of $Q$ into
itself will be said to be a dual poset if, $\forall\, q, r \in Q$ :

1)      $q'' = q$

2)      $q \leq r \Rightarrow r' \leq q'$

3)      $q \neq q'$.

The mapping $q \to q'$ will be called the duality on the dual poset. A
poset $Q$ is said to be bounded if there exists either a greatest element
$1 \in Q$ or a least element $0 \in Q$.  Obviously, then, by 2) and 3) above
a bounded dual poset contains both 1 and 0, and the trivial case with
$1 = 0$ is excluded.  For $Q$ a dual poset, a subset $S \subset Q$ will be said to be

1)      a base for $Q$ if,  $\forall\, q \in Q$, $\{q,q'\} \cap S \neq \emptyset$ ,

2)      dual exclusive if, $\forall\, q \in Q$, $q \in S \Rightarrow q' \notin S$.

Moreover, if $Q$ is a bounded dual poset, $S$ will be said to be

3)      unitary if  $1 \in S$,

4)      an M'-base if  $S$ is a unitary dual-exclusive base for $Q$.

The set of all functions $\varphi : Q \to [0,1]$ on bounded dual poset $Q$ such that:

a)      $\forall\, q, r \in Q$, $q \leq r \Rightarrow \varphi(q) \leq \varphi(r)$

b)      $\varphi(1) = 1$ and $\varphi(q') = 1 - \varphi(q)$, $\forall\, q \in Q$,

will be called the set of probability measures on $Q$, and symbolized by
$\Phi_Q$ .  The definitions for the various types of extensions to be considered for bounded dual posets are those given in D5 of I with "bounded
dual poset" everywhere replacing "orthoposet," "duality" replacing

"orthocomplementation," and the _convention_ that if Q is in fact an ortholattice as defined in I, then duality will always mean the duality provided by the natural orthocomplementation on Q.     □

We note that the M-bases defined in I form a subclass of the M'-bases defined above, since the relation $q \perp r \equiv q \leq r'$ makes sense for dual posets and retains enough of its properties so that is is easily checked that maximal $\perp$ - exclusivit y implies maximal dual-exclusivity.

We can now give the basic "metatheorem" relating I to the present paper:

<u>MT</u>.  All the lemmas, theorems and proofs in I concerning arbitrary orthoposets and their probability measures remain valid for arbitrary bounded dual posets and the probability measures defined on them. Moreover, M-bases may be generalized to M'-bases in all relevant constructions.     □

We leave the straightforward but somewhat lengthy checking of MT to the interested reader.

There is one further slight generalization that can be made, since every unbounded dual poset Q with set of order-preserving functions $\Phi_Q$ uniquely defines a bounded dual poset $\overline{Q}$ with corresponding probability measures $\Phi_{\overline{Q}}$ through the simple expediency of adding a greatest element 1 and a least element 0 to Q to get $\overline{Q}$, and then extending those functions $\varphi \in \Phi_Q$ with the property $\varphi(q') = 1 - \varphi(q)$ to get the measures $\overline{\varphi} \in \Phi_{\overline{Q}}$ by the rule $\overline{\varphi} \mid Q = \varphi$, $\overline{\varphi}(1) = 1$, $\overline{\varphi}(0) = 0$. Hence, we can give a final theorem summing up explicitly the results of I and the present paper:

T.  For Q a dual poset, let $\overline{Q}$ be the bounded dual poset corresponding to Q, and let $\Phi$ be the set of probability measures on $\overline{Q}$. Then for S any maximal unitary dual-exclusive subset of $\overline{Q}$, there exists a Hilbert space H and two mappings $\pi : Q \to P$, $\eta : \Phi \to P_1$, where P is the set of projection operators on H and $P_1 \subset P$ is the set of one-dimensional projection operators on H, such that:

1)    $\pi(1) = 1$

2)    $\forall q \in \overline{Q}, \quad \pi(q') = 1 - \pi(q),$

3)    $\forall q, r \in \overline{Q}, \quad \pi(q)\,\pi(r) = \pi(r)\,\pi(q),$

4)    $\forall q, r \in S, \quad q \leq r \Leftrightarrow \pi(q) \leq \pi(r),$

5)    $\forall \varphi, \psi \in \Phi, \quad \varphi \neq \psi \Rightarrow \eta(\varphi) \perp \eta(\psi),$

6)    $\forall \varphi \in \Phi, \; q \in Q, \quad \varphi(q) = \text{trace}\,(\eta(\varphi)\,\pi(q)). \qquad \square$

Proof: From MT above, we can apply T1 of I to get a Boolean M-ex-
tension of Q with respect to the M'-base S, and use the construction
in the proof of T2 of I to generate a pure QM extension. Conditions
1) through 6) of the present theorem simply spell out in explicit detail
the properties of any extension resulting from that particular method
of construction.        Q.E.D.


We note that T above can be taken as a full summary of the results
of I extended to dual posets, since it implies the existence of measure-
preserving embeddings into Boolean algebras -- we simply restrict
the range of the mapping $\pi$ above to the maximal abelian subalgebra B
of projection operators generated by the set $\pi(Q)$.

3. <u>Conclusions</u>

The only conclusions to be added to those already stated in
I involve the apparent minimality of conditions needed to guarantee a
quantum theoretical model for physical reality. If Q is taken as a set
of physical questions, with each $q \in Q$ understood as a list of operational prescriptions for arriving at a yes-no answer, then there is
already present a natural mapping $q \to q'$ of Q into itself : q' is the
same list of prescriptions as q with only yes-no interchanged at the end.
This mapping obviously satisfies conditions 1) and 3) in the definition
of duality given earlier, i.e., $q'' = q$ and $q' \neq q$, and it gives  enough
structure to Q so that we can select maximal subsets $S \subset Q$ (using
Zorn's lemma if Q is infinite) in analogy with the M'-bases defined
earlier. Given any  such maximal subset S, all that is further
required to specify what is usually called a physical system is a set $\Phi$
of functions $\varphi: S \to [0,1]$ (each function $\varphi$ presumably derived from

experiments involving the physical questions in Q and expressing the
probability of getting a "yes" answer in an empirically prescribed set
of circumstances) satisfying only some very minimal requirement --
for example,

A.        $\forall q \in S, \quad \exists \varphi \in \Phi : \varphi(q) \neq \frac{1}{2}$.

Then, since no new physical operations are involved in q' other than
those already involved in q (only the purely mental operation of exchang-
ing "yes" and "no" at the end), we are free to <u>define</u>

$$\forall q \in Q, \; q \in S, \; \varphi(q') = 1 - \varphi(q)$$

without fear of inconsistency.


Now since we generally do not want to distinguish between
physical questions unless they give distinct probability results in some
set of circumstances (i.e., unless they are distinguishable by some
state) we go to a functional representation of Q as a set of functions
on the states.  That is, we define,

$$\forall q \in Q, \; \widetilde{q}(\varphi) = \varphi \in \Phi,$$

and let $\widetilde{Q}$ be the set of functions on $\Phi$ so defined.  Then $\widetilde{Q}$, under the
natural ordering of functions, is a dual poset with the mapping $q \to q'$
($\equiv 1 - \widetilde{q}$) as duality (A above guarantees that $\widetilde{q} = \widetilde{q'}$ ), and so may be
embedded in a measure-preserving way into an ortholattice of Hilbert
space projection operators once we have chosen an M'-base in $\widetilde{Q}$.
[Note that the functional representation $\widetilde{S}$ of the set S above may fail
to be such a base. ]

   The points to be made from this rather lengthy excursion are:
1)   It may be the best we can do, in view of the counterexamples
     referred to in I, showing that we are not completely free to
     require preservation of probability measures <u>and</u> full
     preservation of ordering or orthogonality.
2)   The worst misbehavior that can occur in the final quantum
     theoretical model is limited to lack of order-preservation
     between elements in the M'-base chosen and its complemen-
     tary set, and we could hope that a base could be chosen so
     that all "physically relevant" ordering would be confined either

to the base or its complement, where it would be preserved. If not, the maximum price to be paid is the presence of some unphysical questions and states in the final model, and these are already present in models with superselection rules.

3)  Nowhere were we obliged to require of physical reality that it in some sense be orthomodular, have meets or joins, or even have a full orthocomplementation, although all these features are present in the final mathematical model.

This suggests the possibility that, as has already been pointed out by Gudder[3] with reference to the hidden variable controversy, many of the properties for which physical motivations are often sought may be solely properties of the mathematical language within which we choose to work.  Whether or not reality has these properties is usually the type of question that is extremely difficult even to formulate much less to verify or falsify, and may be largely irrelevant to the success of a particular model.

A final conclusion might be that the real assumptions of quantum theory do not lie in the area of the existence of Hilbert space models with sufficient power to deal with arbitrary physical situations, but only in the area of assignments of specific operators on a Hilbert space to represent specific observables and states.  The quantum theoretical language itself is flexible enough to serve all necessary purposes, at least until a simpler, more elegant or computationally more tractable language is found.

Acknowledgments.

We wish to thank R. J. Greechie for pointing out the possibility of a generalized theorem, and the Research Corporation for a grant under which this work was done.

## References


1. Private communication to the author.
2. A. R. Marlow, J. Math. Phys. 19, 9, 1841-1846 (1978)
3. S. P. Gudder, Rev. Mod. Phys. 40, 229-231 (1968)

# QUANTUM LOGIC AND QUANTUM MAPPINGS

David Finkelstein[1]

School of Physics
Georgia Institute of Technology
Atlanta, Georgia

Logic quantization and the problem of constructing a
quantum set theory are considered.  As a way of making physical theories, logic quantization is compatible with and
suggests principles of atomism, anarchy, and asymmetry, adopted for further exploration.  The only process of classical set
theory that seems problematical is the exponential, which requires the construction from two quanta X and Y of a new
quantum $Y^X$ or hom(X,Y), the generic mapping from X to Y.  The
obvious quantum analogue is the linear space of linear maps
from the linear space of X to that of Y, but this violates the
classical arithmetic law $Y^{X+X'} = Y^X \cdot Y^{X'}$.  There is a way of
looking at quantum maps that makes this departure of quantum
from classical arithmetic seem an inevitable consequence of
unlimited superposition, and makes intuitively acceptable the
isomorphism of the quantum exponential $Y^X$ with the product XY.

---

[1]Supported by the National Science Foundation.



Copyright © 1980 by Academic Press, Inc.
All rights of reproduction in any form reserved.
ISBN 0-12-473260-7

## I.   PHILOSOPHICAL INTRODUCTION

In this section I select and emphasize some of the ideas
already set forth at this conference, especially by Wheeler
and Marlow, and apologize for any distortions that creep in.

Wheeler likens time space, which in our first encounter
seems totally continuous, symmetric, and deterministic, to a
quantum solid, which is more or less atomic, asymmetric, and
unlawful.  This suggests three general principles I provision-
ally adopt:


1.   ATOMISM.   ALL CONTINUITY IS STATISTICAL.   The emer-
gence of approximate continuity from atomistic foundations is a
familiar idea both in classical and quantum phyiscs, and needs
no explanation here.  The application to time space has often
been attempted, with little success.  The main new insights
afforded by quantum theory are Snyder's that atomism is compatible with continuous symmetry in a quantum theory, Weizsäcker's
that a quantum binary decision process might generate the
Minkowski time space, and K. Wilson's that the details of a
classical lattice structure are lost near critical points and
a continuum approximation becomes exceptionally good.  The
general idea is that we see a continuum because of our gross-
ness, but when one particle looks at another it may see something quite discrete.

Atomicity may also give rise to complementarity effects.
Gödel's theorem, it is well known, is closely related to the
unsolvability of the halt problem for computers.  More generally, I suppose, there may be a deep connection between Gödel

undecidability and Bohr indeterminacy.  Both may be consequences of an underlying anti-Solon principle:  We cannot fully know ourself, because to know a system fully requires a much more complex system.  For Bohr, indeed, a man preparing a photon is attempting to know himself.  I have benefitted from continuing discussions of the possibility that quantum logic might be a consequence of complexity theory with Gregory Chaitin.

It seems hard to get rigorous results about this.  The wellknown program that prints out its own description is irrelevant, for that program does not necessarily learn or <u>know</u> anything.

Moreover, the very idea of such a derivation goes against the operational grain of present quantum physics, against Einstein's faith in a non-malicious deity.  ("<u>Raffiniert ...,</u> <u>aber bosehaft nie</u>.")  If a complete world description exists in some Platonic sense, it is unavailable to us, and God plays with marked cards:  He can read the backs, we cannot.

I put this possibility aside for now.  It is too hard and ugly.  I here suppose that quantum logic is basic, and already includes these limits to self-knowledge.

This raises an old question:  Which quantum logic?  If for now we cling to the projective logics of linear spaces, then over which field?  In accord with the atomic doctrine I presently take my quantum wave functions or bras and kets to have integer components and an integer-valued quadratic form $\delta_{mns}$ forming an "integer Hilbert space."  I know that integers are not a field.  But it is easy to see that such a logic with integers is equivalent to a projective geometry over the rational

field, provided with a polarity.  The integers here express a
challenge:  Derive the complex Hilbert space from the integer
one statistically, presumably along with the time continuum
to which the complex imaginary unit is linked by Schrödinger's
equation.  To make atomicity as meaningful as possible, I
assume finiteness as well in this challenge.


     2.  ANARCHY.  ALL LAW IS STATISTICAL.  Here the challenge
is to derive the Hamiltonian operator or the Lagrangian function of the usual quantum theory from a deeper quantum theory
in which anything goes.  A physical theory may ideally be considered to include a language whose well-formed formulas all
describe feasible experiments, so that the well-formed formulas and the meaningful ones coincide.  In most physical theories, nevertheless, some of these experiments will have null
outcome, so that the scope of syntax still will exceed the
realm of the allowed.  Then dynamical or other laws must be
added to eliminate the processes that are syntactically allowed but physically forbidden.  For instance in a Newtonian
celestial mechanics to find a square orbit for the Earth is
meaningful but forbidden.  We here seek a language for which
the well-formed, the meaningful, and the allowed coincide, and
no dynamical law is necessary.

     Quantum theory is already much closer to such lawlessness
than classical mechanics.  In the classical theory almost no
transition is allowed; indeed, only one, if we specify initial
and final times and initial point of phase space.  In the
quantum theory almost every transition is allowed; indeed, all
but those with exactly orthogonal initial and final ket and

bra, in the analogous problem.

It even seems possible to dispense with the measure zero case of exact orthogonality, if desired, without drastically revising quantum theory.  For example, the simplest forbidden transition is between crossed polarizers.  But even this case is only approximately forbidden, on fundamental grounds.  If we consider the quantum zero-point motion of the polarizers, we see that it is physically impossible to maintain them exactly crossed.  The usual selection rule is the idealized limit as certain quantum numbers associated with the observer part of the system approach infinity.  Some of the relations we think to be laws may approximately describe constraints imposed by the experimenter and may relax when we consider the entire system.

The main lawlike relic of classical physics in today's quantum theory is the law of motion, the Hamiltonian or action principle.  I intend to recover the quantum theory of today, with its vestigial separation of classical observer and quantum system, and its Hamiltonian dynamics, from a more holistic quantum theory with no law analogous to Schrödinger's.  It seems basically wrong to assume Hamiltonians.  We should deduce them statistically from microstructure.

In quantum logic of the kind I mean it is more natural to deduce a Hamiltonian than to assume one.  The deduction proceeds along the familiar road opened by Dirac,

$$\Psi \sim \exp(iS/\hbar),$$

between quantum vector $\Psi$ and action functional $S$; but against traffic.  Most people start from an assumed Lagrangian on the right and use the Dirac formula to deduce quantum amplitudes on the left.  If we describe the physical situation in quantum logic language from the start, then we first name processes by their quantum vector $\psi$, and use the Dirac formula to deduce the action in the classical continuum limit, following the road from left to right.  Several one-particle examples of such quantum logical computations leading to Dirac-like Hamiltonians and equations have been worked out.

It helps in this quantum logic program to know that it is not necessary to postulate a formula for probability.  For the following theorem only I distinguish between initial points (kets) and final points (bras) as predicates, using the fact that quantum logic is a tensed logic.  <u>Theorem</u>:  The forbidden transitions between initial and final points (pure cases) determine the linear space of the quantum theory.  The compulsory transitions between initial and final points then determine the metric of the linear space.

This stronger combination of the well known theorems of Wigner and Birkhoff-von Neumann can be combined with the quantum law of large numbers to compute probabilities too, deriving rather than assuming the expectation value principle of quantum theory.  These results permit us to concentrate on the yes-or-no questions and the points of quantum logic, knowing that the rest will follow peacefully.  But they compel the development of a quantum set theory.  Indeed, probability theory is a poor man's set theory, concerned only with the easiest (frequency) predicates about the easiest (very large) sets, a matter of permutations and combinations for the

atomist.  I shall report in II on what I think was the last
impediment to a full quantum set theory.


3.  ASYMMETRY.  ALL SYMMETRY IS STATISTICAL.  Indeed the
idea of a basic exact symmetry is counteroperational, and
assumes another form of divine malice.  Suppose, for example,
we believe in translational symmetry and invariance.  This is
supposed to tell us that an optics experiment goes the same
ways back in Atlanta and here in New Orleans.  I imagine slowly and gently reducing the separation between the experiment
and the experimenter, keeping external agents out.  At first
I see no change in the distribution of outcomes.  What happens,
however, when I, the experimenter, move the experiment so close
that my ear gets in the beam?  Clearly the outcomes change.
This is inevitable for a physical observer.  If we consider
this changing the experiment, then the whole question of translational invariance is begged, loses operational meaning, and
translational invariance merely refers to a "translation of
the whole universe", an idle postulate.  We had better say instead that translational invariance is a limiting or asymptotic
statement for widely separated systems.

Put it differently:  If only relative quantities are physical, thanks to translational invariance, then we should be
able to express our theory in terms of such quantities exclu-
sively.  Then the original translations have nothing to act on,
and so reduce to the identity, and the relative translations
that remain are only approximate invariances.

Thus, a quantization of relative distance in particular
does not contradict the usual translational invariance.

We have kept in quantum theory the bad classical habit of idealizing ourselves, assuming ourselves physical enough to take data but not physical enough to disturb the conditions of the experiment, in spite of all the lessons of relativity and quantum theory.  I am trying to accept as fully as I can the consequence that an observable symmetry is necessarily an approximate one, approaching the usual idealization only in the unphysical limit as certain quantum numbers connected with the observer are imagined to approach infinity.  (This is the same kind of limit we were forced to consider in our search for anarchy, because law is simply a special case of symmetry, that associated with time translation.)  A fundamental and holistic theory should have no exact symmetry, as well as no exact selection rules.  Every symmetry that remains today is thus a challenge and a puzzle.  It helps that the integer unitary group of II.4, the symmetry of our logic now, is smaller than the complex unitary group.  These three principles require us to give more information about the observing aspect of the experiment than does present quantum theory, and to recover the usual quantum theory only in an ideal limit as an observer parameter grows indefinitely in an essentially unphysical way.  They pose three challenges.

4.  TRAUTMAN'S TOWER.  Marlow escorts us to a most beautiful application of fiber bundle theory to physics, and one that forces us to reconsider yet again the question of being and becoming raised in the _Timaeus_.  In Trautman's Tower there is a story for each stage in the development of physics.  On the bottom story physicists work with Aristotelean geometry, with an invariant distinction between the horizontal directions

and the vertical.  We live on the top story, of course.  The
remarkable thing pointed out by Trautman is that all the physicists from the second story up can say the same words to
their benighted predecessors one story down; surely a hint of
what some physicist may shortly be communicating down to us:

"Idiots!  That object isn't a product bundle.  It's only
a factor in a bigger product."

The shifting boundary between factor bundles represents
the shifting boundary between description of state and transformation of state; between being and becoming, in the old
terms.  Our story informs our neighbors downstairs, who believe special relativity, that the 8 dimensional tangent bundle to time space is not a product because there actually is
no distant parallelism; but we general relativists still believe that the 16 dimensional bundle of _its_ tangent vectors is
a product, because any change in a tangent vector to time
space can be invariantly decomposed into a change of location
and a change of components relative to the connection of the
world.

Is there a top to Trautman's Tower?  Or at least a limit?
An end to the process of going from object to transformation
in search of  true objects?

I think we have already passed the top, in that we have
carried out this procedure more often than will be recognized
as meaningful in the physics of tomorrow.  The most striking
difference between the concepts of classical physics and quantum physics is that the objects of classical physics are trans-
formed by other objects outside themselves, while the objects
of quantum physics are their own transformations.

For example, in classical mechanics canonical transformations require a Poisson bracket or symplectic structure to be given in addition to the algebraic; while in quantum mechanics the algebraic structure of the logic itself suffices, and all observables are also transformations from the start.  Moreover, Heisenberg showed, the classical Poisson bracket theory follows statistically from the quantum algebraic commutator theory. There is an unusual fusion of being and becoming in quantum logic that seems to get us off Trautman's Tower at last.

II.  QUANTUM MAPPINGS

1.  QUANTUM SET THEORY.  By a predicate algebra, as I indicated in I.2, I shall mean a set of elements called <u>point predicates</u> or points, with a symmetric binary relation between points called the <u>null</u> <u>relation</u>.  Rather than axiomatize, I construct.  I assume here that points may be represented (non-uniquely) by columns of integers $(\psi^m)$ and the null relation by $\delta_{mn} \phi^m \psi^n = 0$; and turn now to construct higher logic.  (It is understood that the full lattice of predicates is to be constructed via the Galois connection of the null relation).

A systematic way to begin our construction of a total quantum language is to make a quantum analogue of classical set theory.  Since classical set theory has become a socially accepted well-nigh universal language for mathematics, a quantum set theory is apt to be an acceptable universal language for mathematical theories of quanta.  To be sure, the axioms of classical set theory may -- I believe <u>do</u> -- include counter-natural hypotheses, just as those of classical logic and

geometry do, and we must beware of building these flaws into
our quantum theory.  But if we adhere to the basic principles
of relativity and quantum theory, then, even if these talismans
do not completely protect us against our mathematics, we will
at least have made an interesting model, a relativistic quantum theory.

It is remarkable and encouraging that only a dozen or so
principles, depending on the formulation, make up the language
of almost all modern mathematics.  Of these some, like the
Axiom of Choice, the Axiom of Infinity, and the Axiom of
Foundation, are useful only in infinite universes and may be
left aside here.  Do we need to postulate, say, the existence
of so ill-defined an object as "the set of all integers" in
a theory that is supposed to be physical?

Others, like the Axiom of Pairs, are obviously fulfilled
in the quantum models of sets of quanta that have been the
stock in trade of quantum physics since the early days of
quantum field theory and the quantum theory of polyelectronic
atoms, and require only brief mention here.  Thus, the classical operation  of <u>sum</u> (disjoint union) of sets has a
natural analogue in the quantum operation of direct
product of the linear spaces or modules representing the
quantum sets as quantum objects.  The classical <u>power set</u> $2^x$,
the set of all subsets of the set x, corresponds to the exterior algebra of the linear space in quantum theory.  The classical <u>cardinal numbers</u> 0, 1, 2, 3, ... correspond to the quantum objects $\hat{0}$, $\hat{1}$, $\hat{2}$, ... , described usually by linear spaces
of as many dimensions, and here by the free Abelian groups of
as many generators or, equivalently, by these as modules over

the ring of integers.


2.  THE BASIS RULE.  The general principle behind all
these special cases is:  Do to a quantum basis what you would
do to the classical points.  (Indeed the points of a classical
set _are_ a basis for it.)  Let us call this the _basis rule_.
While classical sets have only one basis, quantum sets have
many, and each use of the basis rule requires a proof of basis
independence.


3.  MAPPINGS.  The last obstacle is the Axiom of Images,
that every set has an image under every mapping (or function)
with that set as domain, and the one obdurate concept is that
of mapping, and specifically the mappings from one object X to
a second Y regarded as constituting a third object Z;
$Z = \mathrm{Hom}(X,Y) = Y^X = (X{\to}Y)$ being three synonomous notations for
this construction, the _exponential_.


4.  THE CATEGORY OF FINITE PROPOSITIONAL ALGEBRAS.  The
categorial approach leads swiftest to the problem.  The classical category we begin from has as objects the cardinal numbers 0, 1, ..., each represented by the set consisting of the
preceding ones, if we wish.  The null relation in object X is
$\neq$, inequality.  Its morphisms or mappings $Y^X$ are the relations
or pairings between points of X and Y with the classical function property:  each point of X enters into exactly one pair.
We may represent such a mapping by a matrix whose columns are
numbered by the points of X and whose rows are numbered by the
points of Y, with a 1 in the (x,y) position if the pair (x,y)

is in the relation, and a 0 otherwise.  The function property
says there is exactly one 1 in every column.  The number of
such matrices is indeed the exponential $Y^X$ we learned in elementary school.

The corresponding quantum objects (over the integers as c
numbers) may still be considered to be the cardinals, but we
put circumflexes over them to show they now belong to a different category with different morphisms: $\hat{0}$, $\hat{1}$, ... .  Call a
matrix M  an <u>integer</u> <u>isometry</u> if its elements are integers
and its product with its transpose is scalar:  MTM = $N\underset{\sim}{1}$, with
integer N; and <u>integer</u> <u>unitary</u> if in addition it is square.
The morphisms from X to Y, $Y^X$, are now the integer isometries
from the integer module over X to that over Y, modulo scalar
integer factors.  (That is, we unite integer matrices that are
proportional with arbitrary rational coefficients into one
integer "ray" in the module of the new object $Z = Y^X$, just as
we unite vectors into rays in ordinary quantum theory.)


5.  THE PUZZLE.  The difficulty is that in the quantum
category we have an isomorphism between this $Y^X$ and the product XY.  While multiplication and addition preserve all their
rules when we go from the classical category to the quantum,
exponentiation collapses, and instead of the classical law

$$Y^{X+X'} = Y^X \, Y^{X'}$$

we have the quantum law

$$Y^{X+X'} = Y^X + Y^{X'} \ .$$

We must construct our quantum arithmetic so it reduces to classical arithmetic in suitable limits.  How can 10x10 "reduce" to $10^{10}$, in any limit?

If we forget category algebra and apply the basis rule, we find for $Y^X$ exactly the same matrices of 0's and 1's as in the classical exponential, and so there are the same classical number of them, of course.  Do we reconcile this with the categorial exponential, or must we choose between the two?


6.  THE RESOLUTION.  Let us apply the basis rule more carefully.  This means verifying basis independence, showing that if we replace either the X or Y basis by a new basis, consisting of superpositions of the old basis points, we merely replace the points of $Y^X$ by superpositions.

This raises a question.  How do we superpose the matrices that according to the basis rule make up $Y^X$?

The categorial result includes a definite law of superposition.  The matrices linearly combine just as matrices usually do.  If we adopt this superposition rule together with the basis rule -- and I see no other reasonable one -- then the $Y^X$ matrices given are not all linearly independent.  They may be linearly expressed in terms of the basic matrices of XY, which have just a single 1 in the entire matrix instead of a 1 in each column.

Let us examine these XY basis matrices as mappings $X \rightarrow Y$ more closely.

Evidently they map every basis element of X into 0 except the one that maps into a basis element of Y.  They thus violate the basis mapping rule, do not map an X basis into a Y basis

in the ordinary sense of mapping, but must be regarded as undefined on all basis elements of X but one.  That brings us to
the key point.

In classical set theory we permit mappings to lose points,
map big sets X into little ones Y.  But we do not permit a
mapping to lose any one point in particular.  It must be defined on every point, though it may map two points into one.

In the quantum theory these two possibilities cannot be
separated.  If a mapping sends two vectors into one, there is
a third vector that the mapping sends into 0, and 0 is not a
point of the logic, does not define a point predicate.  This
third vector is merely the difference of the first two, a
consequence of superposition.  On it the point mapping is undefined.

Either we restrict the concept of mapping to those that
lose no points at all, isomorphisms, or we must admit partially defined mappings as the natural concept of mapping for
quantum objects.  In the first case, isomorphisms, the exponential $Y^X$ is 0 unless $X = Y$; and in the second, $Y^X$ is XY.  I
will follow the second way, of course.

Further, we now see how the quantum exponential reduces to
the classical.  Suppose we give 10 apples to 10 children.
When we say there are 10 000 000 000 ways to map 10 apples
into 10 children, we are assuming the apples and children are
classical objects obeying classical set theory, without superposition.  This is safe enough, but only because they consist
of a vast number N of quantum subsystems:  atoms, particles,
or whatever, and if one of these quanta is out of place by a
deBroglie quarter wavelength, superposition fails.  I suppose

all superselection principles, expressing the strongest kind
of symmetry, result from large numbers in this way, as part
of the general thesis that symmetry is statistical.

But then the mappings we are really asking for are not
$10^{10}$ but a small part of

$$(10\ N)^{(10\ N)}.$$

Applying quantum arithmetic, this is just 100 N·N, and the
usual classical answer of 10 000 000 000 is attainable as long
as N·N > 100 000 000, or N > 10 000.  To get the exact classical result rather than this rough inequality would require a
little consideration of the actual physical process of giving
the apples to the children.

The main application of the exponential in classical
physics seems to be field theory.  There X is time space, Y
is the field variable, and Z is the field considered as one
object.  If we simply carry this classical formulation of
field concepts into quantum theory, we will have to deal with
fields that are not defined throughout time space but only on
variable islands.

Some of these matters, even so simple-seeming a one as the
correspondence between the quantum and classical integers,
need more discussion, which is omitted here for reasons of
time.  A fuller account will be published in the International
Journal of Theoretical Physics.

# BOHR-SOMMERFELD QUANTIZATION IN GENERAL RELATIVITY AND OTHER NONLINEAR FIELD AND PARTICLE THEORIES

Robert Hermann[*]

## I. INTRODUCTION

The quantization of General Relativity is one of the foremost challenges to contemporary mathematical physics. Perhaps one should look to the history of particle quantum mechanics and search for some half-way house between the classical and quantum theories, particularly one which exploits the geometric setting of the theory. This paper presents several ideas for a "Bohr-Sommerfeld" approach based on E. Cartan's theory of exterior differential systems [39] and several related geometric physical concepts.

There seem to be four features of Einsteinian gravitational theory that are related to the quantization problem.

a) The Einstein gravitational equations themselves are complicated nonlinear partial differential equations--certainly the most difficult set that occurs in the classical equations of mathematical physics.

---

[*]Supported by NSF Grant MCS78-06000 and Ames Research Center (NASA) Grant NSG-2252.



Copyright © 1980 by Academic Press, Inc.
All rights of reproduction in any form reserved.
ISBN 0-12-473260-7

b)   There is a rich structure of special solutions; indeed,
these turn out to be the key to the successful experimental and
cosmological implications.

c)   The equations themselves admit an infinite dimensional
symmetry group, the group of diffeomorphisms of the underlying
manifold.

d)   The theory is intimately linked to the most sophisticated,
long-studied and in some sense basic geometric structure--the
Riemannian metrics.

Now, quantum mechanics itself grew up in a completely
different mathematical environment--the much simpler finite-
number-of-degrees-of-freedom system of traditional analytical
mechanics, i.e., the harmonic oscillator, hydrogen atom, etc.
There are, of course, general principles available for making
deductions about much more general systems, but detailed
knowledge of the simple systems which make up the complicated
ones (or that serve as a basis for perturbation calculations)
is essential.  Now, there has been considerable difficulty in
extending quantum mechanics to cover relativistic quantum
fields; this has been one of the major obstacles in the way of
a successful elementary particle theory.  These difficulties are
compounded when the Einstein gravitational equations are con-
fronted as nonlinear relativistic fields.

Of course, what the Einstein theory has, and elementary
particle theory so far lacks, is a compelling geometric foundation and intuition.  Notice also that the elementary particle
equations do not have (so far) the rich structure of special
solutions which has been so invaluable in Einsteinian theory.
(It is an encouraging sign that we are beginning to see such
special solutions appear in gauge theories via the "instanton",

"monopole", and "solitary wave" solutions.  Ultimately, one would hope to see these special solutions unified via Lie and Bäcklund theory.)  I would guess that if it were possible to make a mathematically (and physically) successful quantum gravitational theory, it would most plausibly come from exploitation of the rich geometric features of the theory.  Of course, this observation has been made before--the well-known "super-space" theory of John Wheeler and the "twistor" theory of Roger Penrose [35] are ingenious attempts to do precisely this.  However, my personal belief is that we have not done enough to investigate the first "geometric" quantum theory, the Bohr-Sommerfeld "old" quantization theory.

It was amazing to me to learn from Lanczos' book, The Einstein Decade, pp. 113-115 [34] that Einstein himself was the first to discover the real geometric meaning of these conditions, independently of the "separable" coordinate systems in terms of which they were first formulated.  (Surely the most important lesson from differential geometry that Einstein learned--and that physicists have not completely assimilated to this day-- is the methodological importance of thinking in terms indepen- dent of coordinate systems.  If he were working today, he would no doubt revel in the modern theory of differential manifolds, which is the ultimate version of this principle.)  According to Lanczos, Einstein pointed out the correct interpretation in a remark at a lecture of the German Physical Society which was published in its Proceedings.  To my knowledge, Einstein's remark was completely lost and only re-emerged in 1959 in a paper by J. Keller [36].  (Of course, it might have had a folklore status, and been well-known to various insiders.)  I was very struck by this interpretation of Bohr-Sommerfeld and

it has played a key role in my thought about the geometric
meaning of quantum mechanics. (See my book, Vector Bundles in
Mathematical Physics, Vol. II [11].) This remark has also
played into a development of pure mathematics that can be
sampled in the book Geometric Asymptotics by Guillemin and
Sternberg [37].

Here is Einstein's remark translated into modern differential geometric terms. Consider a mechanical system whose
configuration space is a manifold $Q$. Let $X = T^d(Q)$ be the
cotangent bundle to $Q$. (I must assume that the reader is
familiar with modern manifold theory. See [6,38].) Let

$$H: X \to R$$

be a real-valued function, physically to be identified with the
Hamiltonian or total energy. $X$ has one-differential form $\theta$
intrinsically defined on it, called the contact form. $H$
defines a vector field $V_H$ on $X$ via the following relation:

$$dH = V_H \lrcorner \, d\theta \quad . \tag{1.1}$$

The orbits (or integral curves) of $V_H$ are the solutions of
Hamilton's equations; they are curves $t \to x(t)$ in $X$ such
that

$$\frac{dx}{dt} = V_H(x(t)) \quad , \tag{1.2}$$

where $t \to x(t)$ is the tangent vector curve to $x$.

Now, let $S: Q \to R$ be a real-valued function. Its differential $dS$ is a one-differential form on $Q$, which is a cross-
section map: $Q \to T^d(Q) \equiv X$ of the cotangent bundle. The
contact form $\theta$ has the property that

$$dS^*(\theta) \;=\; dS \quad .$$

A curve $t \to S_t$ in the space of all functions is a solution of the Hamilton-Jacobi equation if

$$\frac{\partial}{\partial t} S_t + dS_t^*(H) \;=\; 0 \quad . \tag{1.3}$$

Look for solutions of (1.3) of the form

$$S \;=\; W - Et \quad , \tag{1.4}$$

with $E$ a constant (of course, the <u>total energy</u>) and $W$ a real-valued function: $Q \to R$. It must then satisfy

$$dW^*(H) \;=\; E \quad . \tag{1.5}$$

Now, (1.5) represents a first order nonlinear partial differential equation for $W$. In terms of the usual canonical coordinates $(q,p)$ for $T^d(Q)$, it takes the explicit form

$$H\left(q, \frac{\partial W}{\partial q}\right) \;=\; E \quad . \tag{1.6}$$

Now, in general, there do not exist <u>globally defined</u> solutions $q \to W(q)$ of Eq. (1.5). However, consider the map

$$\psi(q) \;=\; e^{iW(q)/h} \quad , \tag{1.7}$$

where $h$ is a constant. For a given value of constants $E$ and $h$ there might exist a globally defined $\psi$. Intuitively, this means that $W$ can be "multi-valued", with periods which are multiples of $2\pi h$. These are precisely, for seperable, multi-periodic systems, the <u>Bohr-Sommerfeld quantization</u> conditions with $h$ <u>Planck's constant</u>.

One says that a value is <u>quantized</u> if there
is a $C^\infty$ function

$$\psi : Q \to \text{(complex number of absolute value one)}$$

such that locally, about each point of $Q$, there
is a solution $W$ of (1.5) such that (1.7) is
satisfied.

The condition for the existence of $\psi$ can be expressed in
topological form. A solution of the Hamilton-Jacobi equation
is determined, from another point of view, by a cross-section
map

$$\gamma : Q \to T^d(Q)$$

such that

$$\gamma^*(d\theta) = 0 \quad .$$

Hence,

$$d\gamma^*(\theta) = 0 \quad .$$

$\gamma^*(\theta)$ is thus a closed one-form in $Q$, hence (via de Rham's
theorem) determines an element of $H^1(Q,R)$ the <u>first cohomology</u>
<u>of</u> $Q$ <u>with real coefficients.</u> $H^1(Q,Z)$, cohomology with
integer coefficients can also be defined: there is a linear
map

$$H^1(Q,Z) \to H^1(Q,R) \quad .$$

The Bohr-Sommerfeld condition means then that the element of
$H^1(Q,R)$ determined by $\gamma^*(\theta)$ arises from an element of
$H^1(Q,Z)$. In more classical language, this means that $\gamma^*(\theta)$
has integer periods.

This is a crude concept compared to the sophisticated "wave mechanical" formalism that followed.  It is well-known that Einstein was, to the end, implacably opposed to the post-1925 quantum theory.  The intuitive ideas and the mathematical possibilities of the Bohr-Sommerfeld theory must have exerted a continual attraction on his thoughts.  He continually expressed his belief in an ultimate theory which would involve classical fields, certainly without "probability", with something like "quantization" conditions arising from the condition for the global solutions of nonlinear field equations.

I believe that Einstein's intuition here is not misguided, that there might be possibilities of utilization of ideas of "global" differential geometry for purposes of quantization of fields.  My purpose here is to outline certain preliminary ideas which might point in this direction.  The main idea is to utilize a formalism which involves writing calculus of variations problems in terms of differential forms [4,10].  A variational problem of $m$ independent variables gives rise to forms of degree $m$, hence to "quantization" conditions involving $m$-th degree cohomology groups.  In physical field-theoretic situations, $m = 4$.

Here is the very simple, but general, formulation of these ideas.  Let $X$ be a manifold.  An <u>exterior differential system</u> [3,39] on $X$ is a collection of differential forms which is closed under exterior differentiation and exterior multiplication.  Suppose that $\mathscr{E}$ is such an exterior system and that $\theta$ is an m-th degree differential form on $X$.  Suppose $\theta_1, \ldots, \theta_n$ are differential forms which generate $\mathscr{E}$.  Form

$$\omega \;=\; \theta + \theta_1 \wedge \lambda_1 + \cdots + \theta_n \wedge \lambda_n \quad,$$

where $\lambda_1, \ldots, \lambda_n$ are arbitrary differential forms.  Now, choose the $\lambda$ so that $d\omega$ takes the following canonical form

$$d\omega = \eta_1 \wedge \theta_1 + \cdots + \eta_n \wedge \theta_n \quad ,$$

where $\eta_1, \ldots, \eta_n$ are other forms on X.  Let $\mathcal{E}'$ be the exterior differential system generated by $\mathcal{E}$ and the $\eta_i$.  The <u>extremals</u> of the variational problem are then the m-dimensional integral submanifolds of the system $\mathcal{E}'$.

Thus, the extremals are also integral submanifolds of $d\omega$.  However, they are not necessarily the maximal integral submanifolds.  One might find submanifolds Y on which $d\omega$ is zero, which are fibered by the extremals.  Such objects are the generalization of extremal fields.  On such a Y, $\omega$ is a closed m-form, hence defines an element of $H^m(Y,R)$, the cohomology of Y with real numbers as coefficients.  One might now require, as a generalization of Bohr-Sommerfeld, that this cohomology class come from an element of $H^m(Y,Z)$, the cohomology with integer coefficients.

Unified field theories are also very much in the air.  Of course, it is open to question whether we will ever see the grand unification of all the fields--gravitational, weak, strong and electromagnetic.  Mathematically, this would correspond to a grandiose scheme of unification of all the "quantizations".  For example, there is a variant of the Bohr-Sommerfeld conditions involved in Dirac's "monopole" quantization, which is formulated in an extremely attractive way by the Kostant-Souriau "quantization" in terms of complex line bundles [37].  There are also links to the Feynman path integral quantization.  In my book, <u>Yang-Mills</u>, <u>Kaluza-Klein and the Einstein Program</u>

[33], I have discussed some geometric ideas which might serve these physical purposes, involving Riemannian metrics in fiber bundles with space-time as base, in the spirit of the 1920 Kaluza-Klein geometrization of electromagnetism.

I will now explore a general setting--the notion of conservation law of an exterior system and certain related topological ideas.

## II.  CONSERVATION LAWS OF EXTERIOR DIFFERENTIAL SYSTEMS AND BOHR-SOMMERFELD GENERALIZATIONS

Let  X  be a manifold and let  $\mathscr{E}$  be an exterior differential system.  A differential form  $\theta$  on  X  is said to be a <u>conservation law</u> for  $\mathscr{E}$ ' if the following condition is satisfied:

$$d\theta \in \mathscr{E} \qquad . \tag{2.1}$$

See [26] for further discussion and description of the relation to the more traditional notion of conservation law.

Suppose an m-form  $\theta$  satisfying (2.1) is given.  Now, the fact that this formulation involves differential forms and that differential forms have (via the de Rham theorems) intimate relations with topology, immediately suggests certain generalizations of the Bohr-Sommerfeld ideas.

Let  Y  and  B  be manifolds with  $\pi: Y \to B$  a fiber space map.  Thus, for  $b \in B$, $\pi^{-1}(b)$  is a submanifold of  Y. Consider a map

$$\phi: Y \to X$$

with the following properties.

a)  $\phi$  is a submanifold map,

b)  $\phi^*(d\theta) = 0$  , $\tag{2.2}$

c)      For each  $b \in B$,  $\phi$  restricted to  $\pi^{-1}(b)$

is an integral submanifold of  $\mathscr{E}$,  i.e., for

$\omega \in \mathscr{E}$,  $\pi^*(\omega)$  is zero when restricted to the

submanifolds  $\pi^{-1}(b)$.
(2.3)

Consider the geometric meaning of these conditions.  (2.3) means that  $b \rightarrow \pi^{-1}(b)$  is a family, parameterized by b, of integral manifolds of  $\mathscr{E}$.  (2.2) means that  $\phi^*(\theta)$  is a closed m-form on  Y,  hence defines an element of  $H^m(Y,R)$,  the  m-th cohomology group of  Y  with _real_ coefficients.

Now, conditions (2.1) and (2.3) guarantee that  $\phi^*(d\theta)$  is zero when restricted to all the submanifolds  $\pi^{-1}(b)$.  Thus, (2.2) is an _additional_ "integrability" condition for the family  $b \rightarrow \pi^{-1}(b)$  of integral manifolds.  Condition (2.2) enables us to discuss a primitive sort of "quantization"; we can do this by prescribing conditions which must be satisfied by the element of  $H^m(Y,R)$  determined by  $\phi^*(\theta)$.  For example, the most immediate generalization of Bohr-Sommerfeld would be to require that this element lie in the image of  $H^m(Y,Z)$  ($Z \equiv$ integers) under the inclusion  $Z \rightarrow R$,  i.e., that  $\theta$  have "integral periods" on the m-cycles of  Y.

To illustrate, consider the analytical mechanics situation where  m = 1  [6,38].  Let  Q  be a manifold and let  $T^*(Q)$  be the cotangent bundle to  Q.  Let  $\theta$  be the contact one-forms on  X  [38].  Let  $H: X \rightarrow R$  be a real-valued function, the total energy of a conservative mechanical system.  For a real number  E,  let

$$X_E = H^{-1}(E)$$

be the "energy surface".  Restrict  $d\theta$  to  $X_E$,  and let  $\mathscr{E}$  be

the exterior system generated by the one-differential form
which are zero on the Cauchy characteristic curves of $d\theta$
restricted to $X_E$. It then follows readily, using standard
linear algebra, that $d\theta \in \mathscr{E}$, i.e., $\theta$ restricted to $X_E$ is
a "conservation law". (Cartan calls it a "relative integral
invariant" [42].) We can then take $Y = X_E$, with the family
$b \to \pi^{-1}(b)$ of integral manifolds of $\mathscr{E}$ just that provided by
the Cauchy characteristic curves of $d\theta$ restricted to $X_E$.
"Bohr-Sommerfeld quantization" now means that $\theta$ restricted to
$X_E$ belong to

$$H^1(X_E, Z) \quad .$$

To see the relation with traditional Bohr-Sommerfeld, take

$$Q = R$$
with coordinate $q$ .

$T^*(Q)$ then has coordinates $(p,q)$ with

$$\theta = p \, dq \quad .$$

$h$ is a function $H(p,q)$, i.e., a Hamiltonian in the usual
sense

$$X_E = \{(p,q): H(p,q) = h\} \quad .$$

The "Bohr-Sommerfeld" condition is then that

$$\int_{X_E} p \, dq = \text{integer} \quad .$$

i.e., the usual one in the physics books (with Planck's
constant suitably normalized).

We see that the theory of exterior differential systems
gives us a vast freedom of formulating broader sorts of Bohr-
Sommerfeld conditions!

### III.   JET BUNDLES AND THE CONTACT SYSTEM

The usual variational principle describing Einsteinian
gravity can be formulated as follows.  Let  $x$  denote a point
of space-time, i.e.,  $R^4$.  Consider metric tensors as tensorial
functions  $x \rightarrow g(x)$  of  $x$.  (In terms introduced below, they
are cross-sections of a fiber space.)  The scalar curvature  $K$
is a fourth degree differential form on  $R^4$  whose coefficients
are functions of  $g$  and its first and second order derivatives

$$K(g, g_x, g_{xx})\, dx \qquad .$$

The <u>action</u> associated to a metric  $x \rightarrow g(x)$  is

$$\int_{R^4} K\, dx \qquad .$$

The field equations are obtained by extremizing this integral.

I will now describe a general technique for handling
problems of this sort.  Let  $Y$  and  $X$  be manifolds,  $\pi: Y \rightarrow X$
a map.  (All data is, unless mentioned otherwise, of differentiability class  $C^\infty$.)

$\mathscr{F}(X)$  denotes the algebra of  $C^\infty$,  real-valued functions.
For  $x \in X$,  $X_x$  denotes the <u>tangent space</u>.  $T(X)$  denotes the
<u>tangent vector bundle</u>.  ($X_x$  is the fiber of  $T(X)$  above the
point  $x$.)  $\mathscr{V}(X)$  denotes the <u>vector fields</u>, the cross-sections
of  $T(X)$.  $\mathscr{D}^r(X)$  denotes the r-th degree differential forms
on  $X$.  $\pi_*: T(Y) \rightarrow T(X)$  denotes the linear mapping  $\pi$  induces

on tangent vectors.  $\pi$  is said to be a <u>fiber map</u> (or a <u>submersion</u>) if

$$\pi_*(T(Y)) \;=\; T(X) \quad .$$

Let us suppose that this condition is satisfied.

A cross-section map is a map  $\gamma: X \to Y$  such that

$$\pi\gamma \;=\; \text{identity} \quad .$$

Let  $\Gamma$  denote the space of these cross-section maps.

Suppose  $x \in X$,  and  $\gamma_1, \gamma_2 \in \Gamma$.  We say that  $\gamma_1$  and  $\gamma_2$  <u>meet to the r-th order at</u>  $x$  if  $\gamma_1(x) = \gamma_2(x)$  and if, in local coordinates, all derivatives of order  $\leq r$  of  $\gamma_1$  and  $\gamma_2$  agree at  $x$.

REMARK.  Here is a more intrinisic way to define this concept. $\gamma_1$  and  $\gamma_2$  meet at  $x$  to the first order if  $\gamma_1(x) = \gamma_2(x)$ and if  $\gamma_{1*}, \gamma_{2*}$,  considered as maps  $X_x \to U_{\gamma 1(x)}$,  agree. Proceed by induction.  $\gamma_1, \gamma_2$  agree to the r-th order if  $\gamma_{1*}$ and  $\gamma_{2*}$  agree to the (r-1)-st order.

DEFINITION.  $^r Y$,  the r-th order <u>prolongation</u> of the fiber space  $Y \to X$  (or the r-th order <u>jet bundle</u>, also denoted as $J^r(Y)$) is defined as follows:  Say that two points  $(x,\gamma)$, $(x',\gamma')$  of  $X \times \Gamma$  are equivalent if  $x = x'$  and if  $\gamma,\gamma'$ agree to the r-th order at  $x$.  $^r Y$  is a fiber space over  $X$; assign to  $(x,\gamma)$  the point  $x$.  Any  $\gamma \in \Gamma$  defines a cross-section  $\partial^r \gamma: X \to {}^r Y$.  For  $x \in X$,  $\partial^r \gamma(x)$  is the equivalence class in  $^r Y$  containing the point  $(x,\gamma)$.

We can now define another fiber bundle over  $^r Y$,  called the <u>bundle of Lagrangians</u>, denoted as  $LA(^r Y)$.  Given a point

$^r y \in {}^r Y$, sitting above a point $x \in X$, the fiber of $LA({}^r Y)$ is the space of n-covectors to $X$ at $x$ (n = dimension $X$.) A <u>Lagrangian</u> $L$ is a cross-section of this bundle. Thus, for each $\gamma \in \Gamma$, $\partial^r \gamma(x)$ is an element of $^r Y$. $L(\partial^r \gamma(x))$ is then an n-covector at $x$. As $x$ varies, we obtain an n <u>differential form</u> $x \to L(\partial^r \gamma(x))$ on $X$. If $X$ is orientable, this form can be integrated over $X$, obtaining a number

$$L(\gamma) \quad = \quad \int_X L(\partial^r \gamma(x)) \quad .$$

It is this real-valued function $\gamma \to L(\tau)$ whose extremals are to be found.

This elegant coordinate-free method must be supplemented by a formalism which allows calculations. We will use Cartan's method of the moving frame for this. The key to this method is to use bases of differential forms for the manifolds which are adapted to the structures involved. In special cases, physicists know these bases by the name "Vierbein".

Let $(\omega^i)$, $1 \le i,j \le n$, be a basis for one-forms on $X$. Since $\pi$ is a submersion, the one-forms $\pi^*(\omega^i)$ on $Y$ are linearly independent; however, they do not form a basis. Let $(\omega^a)$, $n+1, \le a,b \le m = \dim Y$ be another set of one-forms on $Y$ such that

$$(\pi^*(\omega^1), \ldots, \pi^*(\omega^n), \omega^{n+1}, \ldots, \omega^n)$$

form a basis of one-forms on $Y$.

If $\gamma: X \to Y$ is a cross-section map, then

$$\gamma^* \pi^*(\omega^i) \quad = \quad (\pi \gamma)^*(\omega^i)$$

$$= \quad \omega^i \quad .$$

Hence, there are relations of the form:

$$\gamma^*(\omega^a) = \gamma^a_i \omega^i \quad . \tag{3.1}$$

We can also define functions

$$\gamma^a_{ij}, \gamma^a_{ijk}, \cdots$$

by differentiating

$$d\gamma^a_i = \gamma^a_{ij}\omega^j$$

$$d\gamma^a_{ij} = \gamma^a_{ijk}\omega^k \tag{3.2}$$

$$\vdots$$

These functions live on the base space $X$, but depend on the cross-section and the choice of the moving frame. Let us lift them to $X \times \Gamma$ in the following way: Let $y^a_i, y^a_{ij}, \cdots$ be the real-valued functions $X \times \Gamma \to R$ defined as follows

$$y^a_i(x,\gamma) = y^a_i(x)$$

$$y^a_{ij}(x,\gamma) = y^a_{ij}(x) \tag{3.3}$$

and so forth .

Notice that two cross-sections $\gamma$ and $\gamma'$ agree to r-th order at $x$ if and only if

$$\gamma(x) = \gamma'(x)$$

$$y^{ai}(x,\gamma) = y^{ai}(x,\gamma')$$

$$\vdots$$

$$y^a_{i\cdots i_r}(x,\gamma) = y^a_{i_1\cdots i_r}(x,\gamma') \quad .$$

Hence, these functions $f\binom{(\ )}{(\ )}$ pass to the quotient, and live on $^rY$. We shall also consider the one-forms $\omega^i, \omega^a$ as defined on $^rY$. Then, we can define one-forms on $^rY$ as follows

$$\theta^a = \omega^a - y^a_i \omega^i$$

$$\theta^a_i = dy^a_i - y^a_{ij} \omega^j \tag{3.4}$$

$$\vdots$$

$$\theta^a_{i_1 \cdots i_{r-1}} = dy^a_{i_1 \cdots i_{r-1}} - y^a_{i_1 \cdots i_r} \omega^{i_r} \quad .$$

The $\theta$'s are called the <u>contact forms</u>. Let $\mathscr{C}$ be the exterior differential system generated by the $\theta$'s, i.e., the smallest set of differential forms on $^rY$ which is closed under exterior derivative and exterior multiplication. $\mathscr{C}$ is called the <u>contact system</u>, and plays a particularly important role in the work of Lie and Cartan. They have the following property:

> If $\gamma: X \to Y$ is a cross-section map, and if $\partial^r\gamma: X \to {}^rY$ is its prolongation, then $\partial^r\gamma$ is an integral manifold of the exterior system $\mathscr{C}$, i.e.,
>
> $$(\partial^r\gamma)^*(\theta) = 0$$
>
> for all $\theta \in \mathscr{C}$ .

What is most important for the calculus of variations is that:

> The one-forms $\omega^i$
>
> $$\theta^a, \theta^a_i, \ldots, \theta^a_{i \cdots i_{r-1}}, \ dy^a_{i_1 \cdots i_r}$$
>
> form a basis of one-forms on $^rY$.

Often, one is given the fiber space $Y$ as the product $X \times Z$, where $X$ is a manifold with a coordinate system $(x^i)$, $Z$ with coordinate system $(z^a)$. $Y \to X \times Z$ thus has a coordinate system $(x^i, z^a)$. Set

$$\omega^i = dx^i \; ; \qquad \omega^a = dz^a \quad .$$

Thus, ${}^rY$ has coordinate system

$$(x^i, z^a, y^a_i, \ldots, y^a_{ij}, \ldots) \quad ,$$

with the contact forms defined as follows

$$\theta^a = dz^a - y^a_i dx^i$$

$$\theta^a_i = dy^a_i - y^a_{ij} dx^j \tag{3.5}$$

$$\vdots$$

Often, we work with these coordinate systems--with their use, formulas usually take the simplest and most classical form. Particularly, the exterior derivatives, which are the keys to a successful calculus of variations, are most readily calculable in this form:

$$d\theta^a = - dy^a_i \wedge dx^i$$

$$= - (\theta^a_i + y^a_{ij} dx^j) \wedge dx^j$$

$$= dx^i \wedge \theta^a_i \tag{3.6}$$

(since $y^a_{ij} = y^a_{ji}$). Similarly,

$$d\theta_i^a = dx^j \wedge \theta_{ij}^a \tag{3.7}$$

$$\vdots$$

$$d\theta_{i_1 \cdots i_{r-1}}^a = dx^j \wedge dy_{i_1 \cdots i_{r-1}j}^a \tag{3.8}$$

These formulas can be immediately translated into physicist's lingo.  Think of  X  as the space-time manifold,  Z  as the "field variable".   $z^a, y^a, \ldots$  are the <u>field variables</u> and their derivatives.


IV.   HIGHER ORDER LAGRANGIAN FIELD THEORIES

For simplicity, consider  $^rY$,  locally, as the space of the variables

$$\left( x^i, z^a, y_i^a, \ldots, y_{i_1 \cdots i_r}^a \right) \quad ,$$

$$1 \le i, i_1 \cdots i_r \le n$$

$$1 \le a,b \le m \quad .$$

These variables depend symmetrically on the lower indices. The <u>contact forms</u> are:

$$\theta^a = dz^a - y_i^a \, dx^i$$

$$\theta_i^a = dy_i^a - y_{ij}^a \, dx^j$$

and so on    .

Set

$$dx = dx^1 \wedge \cdots \wedge dx^n \quad .$$

Suppose that $L$ is a real-valued function on $^r Y$. The
_variational problem_ is to extremize

$$\int L \, dx$$

subject to the constraint that the contact forms $\theta^a, \theta_i^a, \ldots$
are zero. To this end, set:

$$\omega = L dx + \theta^a \wedge \lambda_a + \theta_i^a \wedge \lambda_a^i + \cdots + \theta_{i_1 \cdots i_{r-1}}^a$$
$$\wedge \lambda_a^{i_1 \cdots i_{r-1}} \tag{4.1}$$

where $(\lambda_a, \lambda_a^i, \ldots, \lambda_a^{i_1 \cdots i_{r-1}})$ are a new set of differential
forms of the appropriate degree. Our goal is to choose these
forms so that $d\omega$ is in the Grassmann ideal generated by the
contact forms.

Suppose that:

$$dL = L_i dx^i + L_a \theta^a + L_a^i \theta_i^a + \cdots + L_a^{i_1 \cdots i_{r-1}} \theta_{i_1 \cdots i_{r-1}}^a$$
$$+ L_a^{i_1 \cdots i_r} dy_{i_1 \cdots i_r}^a \quad .$$

Hence,

$$d\omega = \left( L_i dx^i + L_a \theta^a + L_a^i \theta_i^a + \cdots + L_a^{i_1 \cdots i_r} dy_{i_1 \cdots i_r}^a \right) \wedge dx$$
$$+ dx^i \wedge \theta_i^a \wedge \lambda_a - \theta^a \wedge d\lambda_a + \cdots + dx^j \wedge dy_{i_1 \cdots i_{r-1} j}^a$$
$$\wedge \lambda_a^{i_1 \cdots i_{r-1}} - \theta_{i_1 \cdots i_{r-1}}^a \wedge d\lambda_a^{i_1 \cdots i_{r-1}} \quad .$$

Thus, if we choose $\lambda_a^{i_1\cdots i_{r-1}}$ so that:

$$dx^j \wedge dy_{i_1\cdots i_r j}^a \wedge \lambda_a^{i_1\cdots i_{r-1}} = L_a^{i_1\cdots i_r} dy_{i_1\cdots i_r}^a \tag{4.2}$$

then

$$d\omega = \theta^a \wedge (L_a dx - d\lambda_a) + \theta_i^a \wedge (L_a^i dx - dx^i \wedge \lambda_a)$$

$$\vdots$$

$$= \theta^a \wedge \eta_a + \theta_i^a \wedge \eta_a^i + \cdots + \theta_{i_1\cdots i_{r-1}}^a \wedge \eta_a^{i_1\cdots i_{r-1}} \tag{4.3}$$

where the $\eta$'s are appropriately chosen forms.  The <u>extremals</u> of the variational problem are then the n-dimensional integral manifolds (on which $dx \neq 0$) of the exterior differential system $\mathscr{E}$ generated by the $\theta$'s and $\eta$'s.  Further details are in [6].

Einsteinian gravitation is a special situation.

$$n = 4 ; \qquad r = 2 \quad . \tag{4.4}$$

$Y$ has coordinates $(x^i, g_{ij})$, with $g_{ij} = g_{ji}$.  (The cross-sections of the fiber space $Y \to X$ are then the space of Riemannian metrics on $X$.)   $^r Y$ then has coordinates

$$(x^i, g_{ij}, g_{ij,k}, g_{ij,k\ell}) \quad .$$

The contact forms are:

$$\theta_{ij} = dg_{ij} - g_{ij,k} \, dx^k$$

$$\theta_{ijk} = dg_{ijk} - g_{ij,k\ell} \, dx^\ell \quad .$$

L  is the complicated expression which is the scalar Riemannian
curvature.  One could then readily work out the four-forms  $\omega$
and the three-forms  $\eta$.  We see then that the differential form
$\omega$  is a <u>conservation law</u> for the exterior system  $\mathscr{E}$.  Thus, we
can use the general formalism sketched in Section 2 in order to
formulate Bohr-Sommerfeld quantization conditions.


V.   THE EXTERIOR DIFFERENTIAL SYSTEM
WHICH DIRECTLY DEFINES THE
EINSTEIN EQUATIONS

Now let us turn to describing the Einstein gravitational
equations directly in terms of Cartan's theory of exterior
differential systems.  Let  X  be a four-dimensional manifold
with local coordinates labelled as

$$(x_i) \quad , \qquad 1 \le i,j,k \le 4, \quad \text{summation convention} \quad .$$

Let  Y  be another manifold with  $\pi: Y \to X$  a fiber space
mapping with  X  as base space.  Suppose that  Y  has local
coordinates

$$(h_{ij}, \gamma_{ijk}, x_i)$$

with

$$\pi(h, \gamma, x) = x \quad .$$

Thus, the coordinates  $(h_{ij}, \gamma_{ijk})$  are coordinates of the
fiber.  In addition, suppose that

$$\gamma_{ijk} + \gamma_{jik} = 0 , \qquad \det(h_{ij}) \neq 0 \quad . \tag{5.1}$$

Set:

$$\omega_i = h_{ij}\, dx_j \tag{5.2}$$

$$\omega_{ij} = \gamma_{ijk} \, dx_k \tag{5.3}$$

$$\theta_i = d\omega_i - \omega_{ij} \wedge \omega_j \tag{5.4}$$

$$\Omega_{ij} = d\omega_{ij} - \omega_{ik} \wedge \omega_{kj} \tag{5.5}$$

$$dx = dx_1 \wedge dx_2 \wedge dx_3 \wedge dx_4 \quad . \tag{5.6}$$

Let $\mathscr{E}$ be the exterior differential system generated by the two-forms $\theta_i$. Let $\phi: X' \to Y$ be a four-dimensional integral submanifold of $\mathscr{E}$, such that

$$\phi^*(dx) \neq 0 \tag{5.7}$$

Then, the pull-backs of $x$ form a local coordinate system for $X'$. Hence, $X'$ can be defined locally by giving $h_{ij}$ and $\gamma_{ijk}$ as functions of $x$. Let

$$Q = \omega_i \cdot \omega_i \quad . \tag{5.8}$$

($\cdot$ = symmetric product of differential forms. $Q$ is a quadratic differential form on the manifold $Z$.) We see that, with condition (5.8) satisfied, $\phi^*(Q)$ is a (positive definite) Riemannian metric on the manifold $X'$. The one-forms $\phi^*(\omega_i)$ are on an <u>orthonormal moving frame</u> for this metric, which is called a <u>tetrad</u> by relativists.

We now see that the condition that $\phi^*(\mathscr{E}) = 0$ requires that

$$d\phi^*(\omega_i) = \phi^*(\omega_{ij}) \wedge \phi^*(\omega_j) \quad . \tag{5.9}$$

This means that the one-forms

$$\phi^*(\omega_{ij}) = \gamma_{ijk}(x)\, dx_k \tag{5.10}$$

are the <u>connection forms</u> for this moving frame, i.e., the
functions $\gamma_{ijk}(x)$ are the <u>Christoffel functions</u> for the
metric terms

$$\phi^*(Q) = g_{ij}(x)\, dx_j \;. \tag{5.11}$$

REMARK.  Here we see why Cartan [43] calls Eqs. (5.9) the
"structure equations" of Riemannian geometry; they are the
equations in the exterior algebra that rigidly determine the
geometric objects.

We have now constructed the exterior differential system
$\mathscr{E}$, so that its four-dimensional integral submanifolds satisfying the "transversality" condition (3.7) are related directly
to the set of <u>all</u> local Riemannian metrics in the variables
$(x_i)$.   (One can prove that, conversely, such a Riemannian
metric $ds^2 = g_{ij}(x)\, dx_i\, dx_j$ determines an integral submanifold
of $\mathscr{E}$.)  In Einstein gravitational theory we must further consider the set of all such metrics whose Ricci tensor is zero.
Now, for a four-dimensional integral manifold $\phi: X' \to Y$
satisfying (5.7), the pull-back forms

$$\phi^*(\Omega_{ij})$$

are the curvature forms.  In order to use Cartan's ideas, we
must express the Ricci terms in terms of differential forms.
In order to do this, let $V_i$ denote the vector fields on $Y$
which are given by the following formula:

$$V_i = h_{ji}^{-1} \frac{\partial}{\partial x_j} \quad , \tag{5.12}$$

where $(h_{ij}^{-1})$ is the inverse matrix to $(h_{ij})$, i.e.,

$$h_{ik}^{-1} h_{kj} = \delta_{ij} \quad . \tag{5.13}$$

Thus,

$$\omega_i(V_j) = h_{ik} dx_k \left( h_{\ell j}^{-1} \frac{\partial}{\partial x_\ell} \right)$$

$$= h_{ik} h_{\ell j}^{-1} \delta_{k\ell}$$

$$= h_i \, h_{\ell j}^{-1}$$

$$= \delta_{ij} \tag{5.14}$$

Geometrically, the $(V_i)$ are the dual vector fields to the moving frame $\omega_i$. Set:

$$R_j = V_i \lrcorner \Omega_{ij} \quad . \tag{5.15}$$

The one-forms $R_j$ are called the _Ricci forms_. Let $\mathscr{E}'$ be the exterior differential system on $Y$ generated by the forms of $\mathscr{E}$ and the Ricci form $R_i$.

THEOREM 3.1. The four-dimensional integral manifolds of $\mathscr{E}'$ on which $dx \neq 0$ are, locally, the four-dimensional positive Riemannian metrics whose Ricci tensor is zero.

PROOF. On such an integral manifold

$$\phi: X' \to Y$$

$$\phi^*(\Omega_{ij}) \;=\; R_{ijk\ell}\,\phi^*(\omega_k \wedge \omega_\ell) \tag{5.16}$$

The functions $(R_{ijk\ell})$ are then the components of the Riemann curvature tensor.  Using (3.14), we see that

$$\phi^*(V_i \lrcorner\, \Omega_{ij}) \;=\; 2R_{iji\ell}\,\phi^*(\omega_\ell) \qquad .$$

$R_{j\ell} \equiv R_{iji\ell}$ is the Ricci tensor.    Q.E.D.

We have now succeeded in constructing an exterior differential system $\mathscr{E}'$ on the manifold $Y$, whose four-dimensional integral submanifolds are the positive definite Riemannian metrics of Ricci tensor zero.

Of course, this is not directly the situation of interest in relativity, since the metrics are positive.  One can readily verify the formulas, either by allowing the $h_{ij}$ to be complex, à la Minkowski, or by replacing the Euclidean by the Lorentz metric.  It just so happens that in Cartan's formalism it is notationally simpler to work with the positive Riemannian case.

VI.   THE EINSTEIN-HILBERT VARIATIONAL
PRINCIPLE IN CARTAN'S FORMALISM

In Section IV, I have shown how general multiple-integral variational problems lead to "conservation laws" for the exterior differential systems whose integral submanifolds are the extremals.  Of course, such an approach does not exploit the special feature of the variational principle whose extremals are the solutions of the Einstein gravitational equations.  I will now briefly sketch a method--which is still under development--for doing this in a more coordinate free way, in parallel with the way the variational principle for Maxwell-Yang-Mills was presented [33].

Let  W  be a manifold.  As before, choose indices  $1 \leq i,j,k \leq 4$  and the summation convention on these indices.  I shall again simplify notation and calculations by using only lower indices and the Euclidean metric tensor.  Let

$$(\omega_i, \omega_{ij})$$

be a collection of one-forms on  W  with

$$\omega_{ij} + \omega_{ji} = 0 \quad .$$

Set:

$$\Omega_{ij} = d\omega_{ij} - \omega_{ik} \wedge \omega_{kj} \qquad (6.1)$$

$$dx = \omega_1 \wedge \omega_2 \wedge \omega_3 \wedge \omega_4 \quad .$$

Let

$$\varepsilon_{ijk\ell}$$

be the $\underline{\text{Levi-Civita tensor}}$, i.e.,

$$\varepsilon_{ijk\ell} = 1 \quad ,$$

and it depends, skew-symmetrically, on the indices.  Set:

$$K = \Omega_{ij} \wedge \varepsilon_{ijk\ell} \, \omega_k \wedge \omega_\ell \qquad (6.2)$$

The $\underline{\text{Einstein-Hilbert}}$ variational problem is to extremize

$$\int K \quad ,$$

subject to the constraints

$$\theta_i \equiv d\omega_i - \omega_{ij} \wedge \omega_j = 0 \qquad (6.3)$$

REMARK.  If the $(\omega_i)$ are orthonormal moving frames for a (positive) Riemannian metric $ds^2 = \omega_i \cdot \omega_i$, then one recognizes (6.2) as the formula for the <u>scalar curvature</u> multiplied by the Riemannian volume element.  (Eq. (6.3) is the constraint which guarantees that $\omega_{ij}$ are the Levi-Civita connection forms, hence that the $\Omega_{ij}$ are really the Riemannian curvature forms.) For notational convenience, we are working with the positive Riemannian case; the <u>formulas</u> we derive will carry over to the non-positive case.  (Or, one can use Minkowski's trick and work with <u>complex</u> tangent vectors.)

In order to set this up by analogy with the way Maxwell-Yang-Mills equations were treated in [33], set

$$\Delta = K - (d\omega_i - \omega_{ij} \wedge \omega_j) \wedge \lambda_i \quad . \tag{6.4}$$

Let us try to choose these forms $\lambda$ so that $d\Delta$ is in the Grassmann ideal generated by the constraints (6.3).  Now,

$$d(\omega_{ij} \wedge \omega_j) = d\omega_{ij} \wedge \omega_j - \omega_{ij} \wedge d\omega_j$$

$$= (\Omega_{ij} + \omega_{ik} \wedge \omega_{kj}) \wedge \omega_j - \omega_{ij} \wedge (\omega_{jk} \wedge \omega_k + \theta_j)$$

$$= \Omega_{ij} \wedge \omega_j - \omega_{ij} \wedge \theta_j \quad .$$

Hence,

$$d\Delta = dK + (\Omega_{ij} \wedge \omega_j - \omega_{ij} \wedge \omega_j) \wedge \lambda_i - \theta_i \wedge d\lambda_i \quad . \tag{6.5}$$

In order to compute $dK$, it is first necessary to compute $d\Omega_{ij}$.  It is readily seen that the following relation holds. (It is basically the classical Bianchi identity.)

$$d\Omega_{ij} = \omega_{ik} \wedge \Omega_{k\ell} - \Omega_{ik} \wedge \omega_{k\ell} \tag{6.6}$$

Using (6.2), we now have:

$$dK = (\omega_{ik} \wedge \Omega_{kj} - \Omega_{ik} \wedge \omega_{kj}) \wedge \varepsilon_{ij\ell m}\,\omega_\ell \wedge \omega_m$$

$$\tag{6.7}$$

$$+\, 2\Omega_{ij} \wedge \varepsilon_{ijk\ell}(\theta_k + \omega_{km} \wedge \omega_m) \qquad .$$

We see that the Lagrange multiplier form $\lambda_i$ can be chosen so that there are forms $\eta_i$ satisfying the following relations

$$d\Delta = \theta_i \wedge \eta_i$$

An <u>extremal</u> will be an n-dimensional submanifold $X$ of $W$ such that:

a) $\quad \theta_i = 0$ on $X$

b) $\quad \eta_i = 0$ on $X$

c) $\quad dx = \omega_1 \wedge \cdots \wedge \omega_n \neq 0$ on $X$.

On this submanifold,

$$\Omega_{ij} = R_{ijk\ell}\,\omega_k \wedge \omega_\ell$$

where $(R_{ijk\ell})$ is the Riemann curvature tensor of the Riemannian metric

$$ds^2 = \omega_i \omega_i \qquad .$$

Of course we know (since the variational problem we have set up is completely equivalent to the usual one in local coordinates) that these conditions will be equivalent to Ricci flatness of the metric. I believe that this can be done directly by showing that the $\eta_i$ involve what I called the Ricci forms $R_i$ in Section 5, but I have not succeeded in proving such a formula directly .

We can state the problem somewhat differently in the context of Section 5; is the differential form

$$\theta_i \wedge R_i$$

the exterior derivative of a conservation law for the exterior differential system $\mathscr{E}'$ defined there?

## VII.  CONCLUSION

Cartan's theory of exterior differential systems provides marvelous conceptual and geometric techniques for handling nonlinear partial differential equations in a coordinate-free way.  It seems best suited to dealing with problems which arise in a geometric context; perhaps for this reason it has not been widely used in physics, in  spite of the fact that it dates from 1895.  It is precisely in the problems of general relativity (and related problems of Maxwell-Yang-Mills fields) that one might look for such applications, since they are the parts of physics which are closest to differential geometry.

As soon as an exterior differential system has a known "conservation law" differential form, it is possible to formulate Bohr-Sommerfeld rules.  A variational principle gives such a conservation law.  There are also other general methods that provide conservation laws.  (For example, for the typical equations of the theory of nonlinear waves, e.g., Korteweg-de Vries and Sine-Gordon, the Bäcklund transformation-inverse scattering structure provides them.)  We see also that this whole circle of ideas is a generalization to <u>field theories</u> of

the material presented in Cartan's book, <u>Lecons sur les
Invariants Integraux</u> [42], which is the basis for recent work
in the application of differential geometric ideas to particle
mechanics [38].  Thus, we see open before us rather broad
possibilities for applying Cartanian geometric concepts to
important physical situations.


## Bibliography


1. R. Hermann, "Some Differential Geometric Aspects of the
   Lagrange Variational Problem", *Illinois J. Math. 6*, 634-73
   (1962).
2. R. Hermann, "Cartan Connections and the Equivalence
   Problems for Geometric Structures", *Contributions to
   Differential Equations 3*, 199-248 (1964).
3. R. Hermann, "E. Cartan's Geometric Theory of Partial
   Differential Equations", *Advances in Math. 1*, 265-317
   (1965).
4. R. Hermann, "The Second Variation for Variational Problems
   in Canonical Form", *Bull. Am. Math. Soc. 71*, 145-148 (1965).
5. R. Hermann, "The Second Variation for Minimal Submanifolds",
   *J. Math. and Mech. 16*, 473-492 (1966).
6. R. Hermann, "Differential Geometry and the Calculus of
   Variations", Second Edition, Math Sci Press, Brookline,
   MA, 1978 (First Edition, Academic Press, New York, 1969).
7. R. Hermann, "Quantum Field Theories with Degenerate
   Lagrangians", *Phys. Rev. 177*,2453 (1969).
8. R. Hermann, "Current Algebra, Sugawara Model and Differential Geometry", *J. Math. Phys. 11*, 1825-1829 (1970).
9. R. Hermann, "Infinite Dimensional Lie Algebra and Current
   Algebra", Proceedings of the 1969 Battelle-Seattle
   Recontres on Mathematical Physics, Springer-Verlag, Berlin,
   1970.
10. R. Hermann, "Lie Algebras and Quantum Mechanics", W.A.
    Benjamin, New York, 1970.
11. R. Hermann, "Vector Bundles in Mathematical Physics",
    Parts I and II, W.A. Benjamin, New York, 1970.
12. R. Hermann, "Lectures on Mathematical Physics", Vol. 1,
    W.A. Benjamin, New York, 1970.
13. R. Hermann, "Spectrum-Generating Algebras in Classical
    Mechanics", I and II, *J. Math. Phys. 13*, 833, 878 (1972).
14. R. Hermann, "Left Invariant, Geodesics and Classical
    Mechanics on Manifolds", *J. Math. Phys. 13*, 460 (1972).
15. R. Hermann, "Lectures in Mathematical Physics", Vol. II,
    W.A. Benjamin, Reading, MA, 1972.
16. R. Hermann, "Geometry, Physics, and Systems", Marcel
    Dekker, New York, 1973.
17. R. Hermann, "Energy-Momentum Tensors", Interdisciplinary
    Mathematics, Vol. IV, Math Sci Press, Brookline, MA, 1973.
18. R. Hermann, "Topics in General Relativity", Interdiscip-
    linary Mathematics, Vol. V, Math Sci Press, Brookline, MA,
    1973.

19. R. Hermann, "Topics in the Mathematics of Quantum Mechanics",
    Interdisciplinary Mathematics, Vol. VI, Math Sci Press,
    Brookline, MA, 1973.
20. R. Hermann, "Geometric Structure Theory of Systems--Control
    Theory and Physics", Part A, Interdisciplinary Mathematics,
    Vol. IX, Math Sci Press, Brookline, MA, 1975.
21. R. Hermann, "Gauge Fields and Cartan-Ehresmann Connections",
    Part A, Interdisciplinary Mathematics, Vol. X, Math Sci
    Press, Brookline, MA, 1975.
22. R. Hermann, "Geodesics of Singular Riemannian Metrics",
    *Bull. Am. Math. Soc. 79*, 780-782 (1973).
23. R. Hermann, "The Pseudopotentials of Estabrook and
    Wahlquist, the Geometry of Solitons, and the Theory of
    Connections", *Phys. Rev. Lett. 36*, 835 (1976).
24. R. Hermann, "The Inverse Scattering Technique of Soliton
    Theory, Lie Algebras, the Quantum Mechanical Poisson-Moyal
    Bracket and the Rotating Rigid Body", *Phys. Rev. Lett. 37*,
    1591 (1976).
25. R. Hermann, "Geometric Structure of Systems--Control
    Theory and Physics", Part B, Interdisciplinary Mathematics,
    Vol. IX, Math Sci Press, Brookline, MA, 1976.
26. R. Hermann, "Geometry of Non-Linear Differential Equations,
    Bäcklund Transformations and Solitons", Part A, Interdis-
    ciplinary Mathematics, Vol. XII, Math Sci Press,
    Brookline, MA, 1976.
27. R. Hermann, "'Modern' Differential Geometry in Elementary
    Particle Physics", VII GIFT Conference on Theoretical
    Physics, Salamonca, Spain, 1977, Proceedings, A. Azcárraga
    (ed.), Springer-Verlag, Lecture Notes in Physics, 1978.
28. R. Hermann, "The Lie-Cartan Geometric Theory of Differen-
    tial Equations and Scattering Theory" (to appear,
    Proceedings of 1977 Park City, Utah, Conference on
    Differential Equations, C. Byrnes, ed.).
29. R. Hermann, "Prolongations, Bäcklund Transformations, and
    Lie Theory as Algorithms for Solving and Understanding
    Nonlinear Differential Equations", in "Solitons in Action",
    K. Lonngren and A. Scott (eds.), Academic Press, New York,
    1978.
30. R. Hermann, "Toda Lattices, Cosymplectic Manifolds,
    Bäcklund Transformations and Kinks", Parts A and B, Math
    Sci Press, Brookline, MA, 1977.
31. R. Hermann, "Quantum and Fermion Differential Geometry",
    Part A, Math Sci Press, Brookline, MA, 1977.
32. R. Hermann, "Applied Differential Geometry and Systems
    Theory", Proceedings of the 1978 Bordeaux Conference on
    Analysis of Systems, C. Lobry (ed.).
33. R. Hermann, "Yang-Mills, Kaluza-Klein and the Einstein
    Program", Math Sci Press, Brookline, MA.
34. C. Lanczos, "The Einstein Decade: 1905-1915", Academic
    Press, 1974.
35. R.O. Wells, Jr., "Complex Manifolds and Mathematical
    Physics", *Bull. Am. Math. Soc. 1*, 296-336 (1979).
36. J. Keller, "Corrected Bohr-Sommerfeld Quantization
    Conditions for Non-Separable Systems", *Annals Phys. 4*,
    180-188.
37. V. Guillemin and S. Sternberg, "Geometric Asymptotics",
    American Mathematical Society, 1978.
38. R. Abraham and J. Marsden, "Foundations of Mechanics",
    W.A. Benjamin, Reading, MA, 1979.

39. E. Cartan, "Les Systèmes Différentielles Exterieures et
    leurs Applications Geometrique", Hermann, Paris, 1946.
40. P. Dedecker, "On the Generalization of Symplectic
    Geometry", Springer Math Lecture Notes 570, Springer-
    Verlag, 1977.
41. G. deRham, "Variétés Différentiables", Hermann, Paris, 1959.
42. E. Cartan, Lecons sur les Invariants Intègraux, Hermann,
    Parix, 1921.
43. E. Cartan, La Géométrie des Espaces de Riemann, Gauthiers-
    Villars, Paris,

# CONCERNING CANONICAL QUANTIZATION OF GRAVITATION THEORY[*]

Arthur Komar

Department of Physics
Yeshiva University
New York, New York

## ABSTRACT

The canonical quantization program for the general theory of relativity is reviewed. It is shown that if this program could be accomplished in such a fashion that the three Dirac constraints which generate the spatial translations are realized by Hermitian operators on a Hilbert space with a positive definite norm, then the fundamental canonical variables, $g_{ij}$ and $p^{ij}$, can have no non-vanishing matrix elements either on the linear manifold of physical states (the constraint hypersurface) or connecting the physical states to any virtual state. It follows that the operator $e^{ij}$ reciprocal to $g_{ij}$ cannot exist and that the remaining Dirac constraint which generates the time translations cannot even be defined.

## I. INTRODUCTION

The Einstein theory of gravitation was defined in terms of a non-singular symmetric tensor field in four dimensions $g_{\mu\nu}(x^{\alpha})$. (Greek indices range from 0 to 3, Latin indices from 1 to 3). The field equations for the 10 independent components of the field may be derived from an action principle which is invariant under arbitrary four-dimensional

---

[*]Supported in part by the National Science Foundation Grant No. PHY79-09405.

127

Copyright © 1980 by Academic Press, Inc.
All rights of reproduction in any form reserved.
ISBN 0-12-473260-7

curvilinear coordinate transformations:

$$(1.1) \qquad \delta \int R \sqrt{g} \, d^4x = 0$$

where R is the Ricci scalar constructed from $g_{\mu\nu}$, and g is the absolute value of the determinant of $g_{\mu\nu}$. The resulting 10 field equations

$$(1.2) \qquad G_{\mu\nu} \equiv R_{\mu\nu} - \tfrac{1}{2} g_{\mu\nu} R = 0$$

($R_{\mu\nu}$ is the Ricci tensor determined by $g_{\mu\nu}$) reflect the symmetries of the theory in the following fashion. Six of the equations (1.2), namely $G_{St} = 0$, are second order in time and propagates the Cauchy data, $g_{\mu\nu}(x^s, x^o = \text{const})$ and $g_{\mu\nu}^{o}(x^s, x^o = \text{const})$, off the initial arbitrarily chosen space-like surface. The remaining four field equations, $G_{\mu}^{o} = o$, do not involve second time derivatives. They place constraints upon the assignment of Cauchy data. Were this not the case, the solution of the field equations (1.2) would be uniquely determined off the initial surface by the Cauchy data and it would not be possible to perform arbitrary coordinate transformations away from the initial surface, for they alter the form of the solution.

In order for such a set of equations to be mutually consistent it is essential that they satisfy four identities which assure that, once the constraining relations are satisfied on the initial surface, the propagation equation will guarantee that they will be satisfied everywhere. It is a consequence of the Bianchi identities of the associated Riemann tensor that the equations (1.2) do satisfy such a set of 4 identities:

$$(1.3) \qquad G_{\mu \, ; \, \alpha}^{\alpha} \equiv 0$$

(The semicolon denotes covariant differention with respect to $g_{\mu\nu}$.) Indeed Einstein originally arrived at these field equations precisely because these identities were an essential demand.

In order to facilitate the construction of the quantum theory which corresponds to the classical theory of Einstein, Dirac[1] developed the Hamiltonian version of general relativity.  In view of the constraining relations between the initial fields and their first time derivatives ("velocities") it is evident that the canonical version of the theory must have sets of relations constraining the fields and their canonically conjugate momenta.  Dirac found that by adding an adroitly chosen divergence to the Einstein Lagrangian of equation (1.1), and defining the canonical momenta in the usual fashion

$$(1.4) \qquad p^{\mu\nu} \equiv \frac{\delta L}{\delta \dot{g}_{\mu\nu}}$$

four of the canonical constraints may be written simply as  the vanishing of four components of the momentum:

$$(1.5) \qquad p^{o\mu} = 0.$$

Forming the Hamiltonian density in the usual fashion

$$(1.6) \qquad \mathcal{H} \equiv p^{\mu\nu} \dot{g}_{\mu\nu} - L$$

it is found to have the form

$$(1.7) \qquad \mathcal{H} = g_{or} e^{rs} \mathcal{H}_s + \frac{1}{\sqrt{g^{oo}}} \mathcal{H}_L$$

where

$$(1.8) \qquad \mathcal{H}_s \equiv -2 g_{sm} p^{mn}{}_{,n} - (g_{sm,n} + g_{sn,m} - g_{mn,s}) p^{mn}$$

and, in natural units,

$$(1.9) \qquad \mathcal{H}_L \equiv \frac{1}{\sqrt{\gamma}} (g_{mn} g_{rs} - \tfrac{1}{2} g_{mr} g_{ns}) p^{mr} p^n + \sqrt{\gamma} \, {}^3R$$

($e^{rs}$ is the reciprocal matrix, $\gamma$ is the determinant of, and ${}^3R$ is the Ricci scalar constructed from, the three-dimensional spatial metric $g_{rs}$).

The dynamical equations are now the usual Hamilton equations of motion

$$(1.10a) \qquad \dot{g}_{\mu\nu} = \frac{\delta \int \mathcal{H}\, d^3x}{\delta p^{\mu\nu}}$$

$$(1.10b) \qquad \dot{p}^{\mu\nu} = - \frac{\delta \int \mathcal{H}\, d^3x}{\delta g_{\mu\nu}}$$

In view of the fact that $g_{o\mu}$ only occur in $\mathcal{H}$ as the coefficients of $\mathcal{H}_S$ and $\mathcal{H}_L$, in order for the constraint equations (1.5) to be preserved in time, it follows from the vanishing of $\dot{p}^{o\mu}$ as given by equation (1.10b)

$$(1.11) \qquad \mathcal{H}_S = 0$$

and

$$(1.12) \qquad \mathcal{H}_L = 0$$

These are an additional 4 constraints which the canonical field variables must satisfy. In fact, if we translate the canonical variables back into the four dimensional notation of the Lagrangian formalism, these four equations are precisely the four Einstein field equations which were the constraints of that formalism: $G^o_\mu = 0$.

In principle, we must determine whether we grow additional constraints by requiring that the higher time derivatives of equation (1.5) vanish. In the Lagrangian formalism we were assured that there were no additional constraints by virtue of the contracted Bianchi identity, equation (1.3). The analogous statement in the Hamiltonian formalism which guarantees that no further constraints occur as we propagate off the initial surface is obtained as follows.

The Hamiltonian formalism which we have thus far developed is defined in a phase space spanned by the basic canonical variables $g_{\mu\nu}(x^s)$

and $p^{\mu\nu}(x^s)$. However, in virtue of the (primary) constraints, equation

(1.5), the canonical variables $g_{\mu o}$ are completely arbitrary. (Their time

evolution may be expressed by the addition of an arbitrary linear combin-

ation of these primary constraints to the Hamiltonian density of equa-

tion (1.7)). Alternatively, we can employ the primary constraints to

confine our considerations to a reduced phase space spanned by the can-

onical variables $g_{mn}(x^s)$ and $p^{mn}(x^s)$. The $g_{o\mu}$ terms in the Hamiltonian

will, from this perspective, merely be four arbitrary functions which re-

flect our freedom to perform arbitrary coordinate transformations. We

shall take this latter point of view in the remainder of this paper.

In this smaller phase space we define the Poisson brackets between

two functionals of the phase space variables $A(g_{mn},p^{mn})$ and $B(g_{mn},p^{mn})$ in

the customary fashion:

$$(1.13) \qquad [A,B] \equiv \int \left( \frac{\delta A}{\delta g_{mn}(\underline{x})} \frac{\delta B}{\delta p^{mn}(\underline{x})} - \frac{\delta A}{\delta p^{mn}(\underline{x})} \frac{\delta B}{\delta g_{mn}(\underline{x})} \right) d^3x.$$

We know that any functional of the canonical variables can be regarded as

the generator of an infinitesimal canonical transformation and that the

Poisson bracket specifies the infinitesimal change generated in the first

functional by the second functional. Employing this fact it is easily

confirmed that the three (secondary) Hamiltonian constraints $\mathcal{H}_s$, equation

(1.8), generate coordinate transformations within the initial hypersur-

face on which the canonical variables are defined, while the fourth con-

straint $\mathcal{H}_L$, equation (1.9), generates the coordinate transformation which

propagates the canonical variables in direction normal to the initial

surface.[1]

We are now in a position to state the property that our system of

constraints is consistent and that we do not grow additional constraints as we iterate the propagation equations off the initial surface.  Since

the propagation of a given Hamiltonian constraint in any direction may be

expressed by the Poisson bracket of that constraint with any of the re-

maining constraints, it is evident that the consistency of the system of

constraints is assured if the Poisson brackets of the constraints can be

expressed as a linear combination of the constraints themselves.  Indeed,

employing equations (1.8), (1.9) and (1.13), we find by direct computa-

tion

(1.14a)
$$[\mathcal{H}_s(\underline{x}),\mathcal{H}_t(\underline{y})] = \delta_{,s}(\underline{x}-\underline{y})\mathcal{H}_t(\underline{x}) + \delta_{,t}(\underline{x}-\underline{y})\mathcal{H}_s(\underline{y})$$

(1.14b)
$$[\mathcal{H}_s(\underline{x}),\mathcal{H}_L(\underline{y})] = \delta_{,s}(\underline{x}-\underline{y})\mathcal{H}_L(\underline{x})$$

(1.14c)
$$[\mathcal{H}_L(\underline{x}),\mathcal{H}_L(\underline{y})] = \delta_{,r}(\underline{x}-\underline{y})[e^{rs}(\underline{x})\mathcal{H}_s(\underline{x}) + e^{rs}(\underline{y})\mathcal{H}_s(\underline{y})].$$

## II.  CANONICAL QUANTIZATION

Having expressed the Einstein general theory of relativity in a canoni-

cal formalism the procedure for quantization would appear to be straight-

forward.  The canonical variables $g_{mn}$ and $p^{mn}$ would be realized by Her-

mitian operators on a linear vector space satisfying the canonical commu-

tation relations

(2.1)
$$[g_{mn}(\underline{x}),p^{rs}(\underline{y})] = \frac{i\hbar}{2}(\delta^r_m\delta^s_n + \delta^s_m\delta^r_n)\delta(\underline{x}-\underline{y}).$$

A factor ordering for the four constraints of equations (1.8) and (1.9)

would have to be found such that they too are expressed as Hermitian opera-

tors, and such that the algebra of equations (1.14) would continue to be

satisfied as operator equations.  The dynamics of the quantum theory

would be given by

(2.1a)
$$\mathcal{H}_s|\psi\rangle = 0$$

$$(2.6) \qquad \langle \varphi | \xi^s{}_{,m} g_{sn} + \xi^s{}_{,n} g_{ms} + \xi^s g_{mn,s} | \psi \rangle = 0$$

Since $\xi(\underline{x})$ is an arbitrary vector field this relation can only be valid provided

$$(2.6) \qquad \langle \varphi | g_{mn}(\underline{x}) | \psi \rangle = 0$$

Treating equation (2.5) similarly, we can also conclude

$$(2.7) \qquad \langle \varphi | p^{mn}(\underline{x}) | \psi \rangle = 0$$

Although this is a rather disasterous result for the interpretation of the theory, one might yet hope that non-vanishing physical matrix elements could be constructed from observables which are non-linear functionals of the canonical variables, since they could connect physical states through virtual states which do not satisfy equations (2.1). However, let us now consider the expression

$$(2.8) \qquad [g_{ij} g_{kl} H(\xi)] = g_{ij}(\xi^s{}_{,k} g_{sl} + \xi^s{}_{,l} g_{sk} + \xi^s g_{kl,s}) =$$

$$+ g_{kl}(\xi^s{}_{,i} g_{sj} + \xi^s{}_{,j} g_{si} + \xi^s g_{ij,s})$$

Treating this relation in the identical manner as we did equation (2.4) we easily obtain

$$(2.9) \qquad \langle \varphi | g_{ij} g_{kl} | \psi \rangle = 0$$

If we now assume that the Hilbert space has a positive definite norm and we interpolate a complete set of intermediate states $|N\rangle$, setting $|\varphi\rangle = |\psi\rangle$ and $k,l = i,j$ in equation (2.9) we find

$$(2.10) \qquad \sum_N |\langle \psi | g_{ij}(\underline{x}) | N \rangle|^2 = 0$$

or equivalently

$$(2.11) \qquad \langle \psi | g_{ij}(\underline{x}) | N \rangle = 0$$

where now only $|\psi\rangle$ need be a physical state.  In a similar fashion it also follows that

$$(2.12) \qquad \langle\psi|p^{ij}(\underline{x})|N\rangle = 0.$$

It is therefore impossible to form any functional of the canonical variables having non-vanishing physical matrix elements.

A corollary of this result is that the operator $e^{rS}$ cannot exist, for if we take the expectation value, in a physical state, of its defining relation

$$(2.13) \qquad \langle\psi|e^{rm}g_{ms}|\psi\rangle = \delta^r_s$$

we find by interpolating a complete set of states, that the left hand side must vanish identically as a consequence of equation (2.11).  Since the definition of fourth constraint, $\mathcal{H}_L$, requires the existence of $e^{rs}$ for the construction of the Ricci scalar, we conclude that in the present circumstance this constraint cannot even be defined!

## III.  CONCLUSION

We have shown that the general theory of relativity cannot be quantized in a canonical fashion by means of Hermitian operators on a positive-definite linear vector space.  The novel and surprising feature of the present paper is that it was sufficient to consider only the comparatively simple spatial constraints in order to obtain a contradiction.  Indeed we have shown elsewhere[2] that whenever a physical theory has constraining relations which generates a non-trivial gauge group, its quantization cannot be realized by Hermitian operators on a positive-definite linear vector space.

If we abandon the Hermicity requirement it is possible to obtain a
consistent factor ordering for the constraints such that the algebra of
equations (1.14) is preserved.[3]  However, how to realize this algebra on
a linear vector space remains an open question.

## BIBLIGRAPHY

1.  P.A.M. Dirac, Phys. Rev. $\underline{114}$, 924 (1959).

2.  A. Komar, Phys. Rev. D, $\underline{19}$, 2908 (1979).

3.  A. Komar, Phys. Rev. D, in press.

# NEW DIRECTIONS IN RELATIVITY
# AND QUANTIZATION OF MANIFOLDS

Phillip E. Parker

Department of Mathematics
Syracuse University
Syracuse, New York

## NEW DIRECTIONS

This first section will be somewhat complementary to some recent surveys of spacetime singularity theory, for example [6,8,20].

Mathematically, singularities were first detected systematically via geodesic incompleteness, as in the Hawking-Penrose theorems.  Thus one process of interest is that of attaching points to fill in the holes.  There are several known ways to do this, but none seem to necessarily result in manifolds.  Two of the best known are the Schmidt b-boundary [19] and the method of Geroch [8].

So suppose we have a smooth spacetime.  By the theorem of Grauert and Remmert [10] every smooth manifold can be embedded as an analytic submanifold of $\mathbb{R}^n$, for $n$ sufficiently large.  Thus every smooth manifold is locally definable by analytic functions.  Now recall that if one makes the physical assumption that quantum numbers determine the entries of the S-matrix, they are analytic [1].  Also one can, at least in principle, construct spacetime from the S-matrix.  The analyticity of the S-matrix led Chew [5] to the (admittedly vague) hypothesis that "nature is analytic".  Putting all this together we are led to suppose that any physically



Copyright © 1980 by Academic Press, Inc.
All rights of reproduction in any form reserved.
ISBN 0-12-473260-7

realistic process of attaching singular points should have as
its result something which is locally definable by (real)
analytic functions:  i.e., a real analytic space as used in
algebraic geometry.  Such things are not necessarily Hausdorff.

According to Hironaka's Theorem [11], using a locally
finite sequence of blow ups one can produce a real analytic
manifold and a proper map down to the analytic space which is
a diffeomorphism except at the singularities of the analytic
space (and has some additional technical properties that
facilitate computations).  Hence, at least for quite a large
class of the methods for attaching singular points, we can
always work on a smooth manifold.  Both the Schmidt and
Geroch constructions are in this class.

Having arrived at a smooth manifold we are encouraged to
let the geometry carry the singular information.  Another
motivation for this view comes from quantum theory and
spcetral geometry.

Let $(X,\beta)$ be a smooth spacetime - X is a smooth manifold
and $\beta$ is a smooth Lorentzian structure on it.  One usually
assumes that the classical observables considered in a given
problem form a Lie algebra which is a Lie subalgebra of the
smooth functions on the cotangent bundle, $C^{\infty}(T^*X)$ with Poisson
brackets.  One associates to the classical phase space $T^*X$
a Hilbert space H, the quantum phase space, and to the classical observables some self-adjoint operators on H, the quantum
observables.  Usually there is some well-defined relation
between the algebras of observables; e.g., both are representations of a given Lie algebra.  The numbers that can be

obtained by the physical measurement of a quantum observable
are precisely those in its spectrum.  Thus in the case of a
single observable

| | | |
|---|---|---|
| spacetime | $(X, \beta)$ | $(X', \beta')$ |
| Hilbert spaces | H | H' |
| operators | A | A' |

one should call the two spacetimes <u>physically</u> <u>equivalent</u> iff
they are isopectral for the observable; i.e., iff spec A =
spec A'.  In the case of several observables one should require isopectrality for all quantum observables corresponding
to a given physical quantity, for each physical quantity.
(Assuming each physical quantity has an observable on both
spacetimes; alternatively one might define <u>physical</u> <u>quantity</u>
this way.)  Thus a physically realistic spacetime should be
a type of isospectral class of conventional spacetimes.  This
approach provides one means of attaining Brans' goal of removing the "absoluteness" from spacetime [3].

So in the sense of quantum theory what is observable is
the geometry:  geometric quantization of the Lorentzian
structure yields the d'Alembertian or Laplace-Beltrami
operator [2,21].  The manifold itself is not only unobserva-
ble, as the examples of Milnor in dim 16 [16] and Vignèras
[24] in dim $\geq$ 2 show, it is not in general even determinable
from observation.

Again we are encouraged to let the geometry carry the
singular information.

Now both paradigms of spacetime singularity, the big bang
and black holes, feature the unboundedness of physical quantities, such as energy and tidal forces respectively, as

one approaches.  Therefore one should consider distributional
(in the sense of L. Schwartz) Lorentzian structures on smooth
manifolds.  Isham [12, p.216] has shown that they must be
considered in a consistent quantization of gravity.  Historically, the use of distributions proveded enormous insight
in the study of singular behavior in many physical phenomena.
Here we get a clarification of ideas and a firm mathematical
foundation for the study of spacetime singularities.

There is now available a theory of distributional geometry [17] which recovers much of smooth differential geometry.
This is covered in detail in the paper so I shall only briefly summarize here.  A key feature is a construction of flows
(based on L.C. Young) for distributional vector fields which
provides geodesics.  It also allows the completion of Keller's
geometrical theory of diffraction.  In gauge theory the objects of interest are distributional connections [e.g.9].
The theory extends General Relativity across spacetime singularities.  Define a spacetime  to be a smooth connected
(paracompact) manifold provided with a distributional
Lorentzian structure, again denoted $(X, \beta)$.  The set of
spacetime singularities is $\Sigma := $ sing supp $\beta$.  If $\Sigma$  is
nonempty then the smooth part of any geodesic which meets
it is incomplete.  Note that Taub-NUT space is nonsingular
by definition.  The complete Carter family [4] extends to
distributional structures on $\mathbb{R}^4$ via the Łojasiewicz division
theorem.  More generally one obtains curvatures, Bianchi
identities etc. with no additional difficulties.  The special
case in which first derivatives have simple jump discontinu-
ities has been extensively treated by Choquet-Bruhat, Israel,
Lichnerowicz, Sokolov and Starobinsky, and Taub ([22]).  My

direct inspiration was Marsden's 1968 paper on generalized mechanics [14].

The only serious problem left here is that coming from products of distributions.

Frequently one thinks of distributions as linear functionals on the space of compactly supported densities. Now the product of two linear functionals is naturally a second degree homogeneous polynomial map. This suggests that one should consider the $C^\infty$ functionals, which would certainly be closed under products. The appropriate notion of $C^\infty$ seems to be that of H. Keller [13] called $C^\infty_\pi$. The work of Michor and its applications, described in the next section, pretty well establish this. We are presently studying this problem.

There is also a rather curious interpretation of distributional Lorentzian structures. If there are some distributional gauge fields on the spacetime and we assume that $WF(\beta) \subseteq WF(\text{gauge fields})$, then this is a weak form of cosmic censorship that also expresses the ultimate dominance of gravity.

Finally there are connections with Fourier Integral Operator theory. One of these is an FIO-theoretic proof of the existence of Hawking radiation in general spacetimes, which I will report elsewhere.

QUANTIZATION

Somewhat in the spirit of Carl Banns' remarks about removing "absolute" manifolds from spacetimes I would like to suggest a method for quantizing manifolds.

Recently Peter Michor has found what seems to be the correct way of making manifolds out of spaces of maps. This

work is in three papers, the first of which is [15].  Let X
be a smooth (finite dim) manifold.  Roughly, what one does
in the case of $C^\infty(X)$ is to put a topology on it so that it
becomes a smooth manifold modeled on $C_c^\infty(X)$:  in fact, regard
$C_c^\infty(X) \subseteq C^\infty(X)$, put the usual function space (or Schwartz)
topology on $C_c^\infty$, transfer it to all cosets, declare them open,
and then take the weak topology this generates on $C^\infty(X)$.
More generally, one carries out a similar process on $C^\infty(X,Y)$
where Y is another (finite dim) manifold.  In this topology,
called $\mathcal{D}^\infty$, $C^\infty(X,Y)$ is paracompact and normal, but not Baire.

Now H. H. Keller has shown [13] that for Hansdorff locally
convex topological vector spaces the most widely used notions
of differentiability coincide at the $C^\infty$ level.  Essentially,
there is a well-defined notion of $f : E \to F$ being $C^\infty$.  Using
this Michor proves:


<u>Theorem 1.</u>  $C^\infty(X,Y)$ is a smooth ($C^\infty$) manifold with a smooth
functorial tangent bundle.  The canonical identification
$C^\infty(X,Y \times Z) \simeq C^\infty(X,Y) \times C^\infty(X,Z)$ is smooth.


<u>Theorem 2.</u>  The space Diff(X) of all diffeomorphisms of  X
is a smooth manifold which is a Lie group:  composition and
inverse are smooth maps.


<u>Theorem 3.</u>  The space Emb(X,Y) of embeddings of  X  into  Y
is a smooth principal Diff(X)-bundle over the quotient space
Emb/Diff.


Theorem 3 has obvious applications in relativity.  (Ask
Wheeler for them!)  E. Binz at Mannheim is working on these.

I would like to tell you about some applications of Theorem 2. For one thing it allows one to rigorize most of the folklore about smooth actions. Also, it is well known that Diff(X) is isomorphic to Sp Diff(T*X), the symplectomorphisms of the total space of the cotangent bundle, and from Michor's theorems one sees that the isomorphism is Lie. Now Diff(X) acts transitively on X and SpDiff(T*X) acts transitively on T*X\0, so we have a classical phase space.

It is well known that X and Y are diffeomorphic iff $C^\infty(X)$ and $C^\infty(Y)$ are isomorphic rings. Not so well known is the theorem of Pursell and Shanks [18] that X is diffeomorphic to Y iff the Lie algebras of compactly supported vector fields are isomorphic, $\mathfrak{X}_c(X) \simeq \mathfrak{X}_c(Y)$. From the proof of Theorem 2 it follows that the Lie algebra of Diff(X) is $\mathfrak{X}_c(X)$, so we have proved

**Theorem 4.** X is diffeomorphic to Y iff Diff(X) $\simeq$ Diff(Y) as Lie groups.

This solves an old problem in differential topology.

As is easily seen $[\mathfrak{X}_c, \mathfrak{X}_c] = \mathfrak{X}_c$. (To verify this, reduce to local coordinates and use the fact that in $\mathbb{R}^n$, $[g\,\partial_i,\, x^i\partial_i] + [\partial_i, x^i g\partial_i] = 2g\partial_i$, where g is smooth.) Then one can argue as in the theorem of Mather and Thurston [e.g.23] to prove

**Theorem 5.** $\text{Diff}_0(X)$ is simple, where the subscript denotes the identity component.

This is of interest in dynamical systems, where one wants to decompose diffeomorphisms into products of Morse-Smale diffeomorphisms.

Now my Theorem 4 says that we can replace $X$ by $\text{Diff}(X)$ in order to study it. Why? Well, one method of quantization is to find irreducible unitary representations of the group of symmetries. Regarding $\text{Diff}(X)$ as the symmetries of $X$, one tries to carry this out. $[\mathfrak{X}_c, \mathfrak{X}_c] = \mathfrak{X}_c$ and Theorem 5 suggest that essentially all of them can be found by the method of orbits [Cf. 21]. (This would be true if also $H^2(\mathfrak{X}_c, \mathbb{R}) = 0$, Lie algebra cohomology.) So one looks at the coadjoint representation of $\text{Diff}_0(X)$ on $\mathfrak{X}_c'$, the distributional vector fields on $X$.

If $\phi \in \text{Diff}(X)$ and $u \in \mathfrak{X}_c$, then $\text{Ad}(\phi)u = \phi_* u$. Hence the coadjoint representation is given by $\text{Ad}'(\phi)\xi = \phi_* \xi$, for $\xi \in \mathfrak{X}_c'$. Thus the orbits of the coadjoint representation of $\text{Diff}_0$ on $\mathfrak{X}_c'$ are just the orbits of the natural induced action of $\text{Diff}_0$ on $\mathfrak{X}_c'$. Let $\xi_0 \in \mathfrak{X}_c'$ and consider the orbit $\text{Diff}_0(\xi_0)$. Each $u \in \mathfrak{X}_c$ defines a vector field $\tilde{u}$ on $\text{Diff}_0(\xi_0)$ via the one-parameter subgroup of $\text{Diff}_0$ generated by $u$. The map $\mathfrak{X}_c \to T_\xi \text{Diff}_0(\xi_0) : u \to \tilde{u}(\xi)$ is linear and surjective (since the action is transitive), and $\tilde{u}(\xi) = \tilde{u}_1(\xi)$ iff $u=u_1 + z$, where $z \in \mathfrak{X}_c$ is such that $\tilde{z}(\xi) = 0$. But this is equivalent to $\langle \xi, [z,u] \rangle = 0$ for all $u \in \mathfrak{X}_c$, so that if $\sigma \in \Omega^2(\text{Diff}_0(\xi_0))$ is defined by

$$\sigma(\xi) \cdot (\tilde{u}(\xi), \tilde{v}(\xi)) := \langle \xi, [u,v] \rangle$$

then $\sigma$ is weakly nondegenerate since vanishing for every $v \in \mathfrak{X}_c$ implies $\tilde{u}(\xi) = 0$. A straightforward calculation using the Jacobi identity shows that $d\sigma = 0$ whence $\sigma$ is a weak symplectic structure. It follows easily that $\sigma$ is $\text{Diff}_0$ - invariant.

Thus, as in the finite dimensional case, the method of orbits is applicable and will produce the desired representations. It remains to be seen if all irreducible unitary representations of $\text{Diff}_0$ can be obtained this way, and to calculate some concrete examples.

Finally, two questions:

1. If $X$ has additional structure (e.g., a Riemannian or Lorentzian metric) how should one take account of this?

2. Can a Killing form be defined on $\mathcal{X}_c$? Formally, for $u, v \in \mathcal{X}_c$, $\text{ad}(u) = \mathcal{L}_u$ and if $e$ denotes the Schwartz kernel of $\mathcal{L}_u \mathcal{L}_v$ then $\text{tr}(\mathcal{L}_u \mathcal{L}_v) = \int_\Delta e$, where $\Delta$ is the diagonal in $X \times X$. If one can make sense of this as, say, an oscillatory integral then $B(u,v) := \text{tr}(\text{ad}(u), \text{ad}(v))$ would be a distribution.

REFERENCES


1.  Barut, A., <u>The Theory of the Scattering Matrix</u>. Macmillan, 1967.
2.  Blattner, R., in Proc. Symp. Pure Math. 26(1973) 147-165.
3.  Brans, C., these Proceedings.
4.  Carter, B., in Les Houches (1972), pp.57-214. Gordon and Breach, 1973.
5.  Chew, G., <u>S-matrix Theory</u>.
6.  Dodson, C., Int. J. Theor. Phys. 17(1978) 389-504.
7.  Ellis, G., and B. Schmidt, Gen. Rel. Grav. 8(1977)915-953.
8.  Geroch, R., in <u>Relativity</u> ed. by M. Carmeli, S. Fickler and L. Whitten, pp. 259-291. Plenum, 1970.
9.  Glimm, J., and A. Jaffe, Harvard Preprint (1977).
10. Grauert, H., Ann. Math. 68(1958) 460-472.
11. Hironaka, H., Ann. Math. 79(1964) 109-326.
12. Isham, C., J. Roy. Soc. Lond. A 351(1976) 209-232.
13. Keller, H., <u>Differential Calculus in Locally Convex Spaces</u>. Springer (LNM417), 1974.
14. Marsden, J., Arch. Rat. Mech. Anal. 28(1968) 323-361.
15. Michor, P., Cahiers Top. Geom. Diff. 19(1978)47-78,et seq.

16. Milnor, J., Proc. Nat. Acad. Sci. 51(1964)542.
17. Parker, P.E., J. Math. Phys. 20(1979).
18. Pursell, L., and M. Shanks, Proc. A.M.S. 5(1954) 468-472.
19. Schmidt, B., Gen. Rel. Grav. 1(1971) 269-280.
20. Seifert, H.-J., in Springer LNM 570, pp. 539-565.
21. Simms, D., and N. Woodhouse, Lectures on Geometric
    Quantization. Springer (LNP 53), 1976.
22. Taub, A., Berkeley Preprint (1978).
23. Thurston, W., Bull, A.M.S. 80(1974) 304-307.
24. Vignéras, M.-F., Hirzebruch Arbeitstagung (1979)

# THE ORIGIN OF MASS OF ELEMENTARY PARTICLES

Lutz Castell

Max-Planck-Institut zur Erforschung
der Lebensbedingungen der wissenschaftlich-technischen Welt
Starnberg, Germany

## I. INTRODUCTION

The origin of mass of elementary particles is in my opinion one of the basic problems of particle physics, and is probably closely linked with the origin of charge-like symmetries. The experimental results of "large transverse momentum physics"[1] have shown that massive particles and mass zero particles exhibit quite a different structure at high-energy momentum processes ($s, -t, -u \gg \sum_i m_i^2$). So the fact that the rest masses $m_i$ are different from zero (not their actual value) plays an essential role in the high energy domain. This was predicted in 1970 on the basis of a conformal invariant scattering analysis[2], and was later studied in the framework of the quark model[3].

If one takes conformal symmetry seriously the question arises which metric in the conformal compactified Minkowski space is physical. Dirac[4], Segal[5] and Castell[6] chose the metric of the static Einstein space as the globally significant



Copyright © 1980 by Academic Press, Inc.
All rights of reproduction in any form reserved.
ISBN 0-12-473260-7

one. In the spatially closed universe, characterized by the radius  R  the lowest energy state is $\hbar c/R$, and therefore the average number of photons must be finite. In the approximation of no interaction  Bose-Einstein condensation of the 2.7 K cosmic photon radiation must take place[7], and there exists a "liquid" phase of photons besides  the usual photon vapour. So the question arises:  are massive elementary particles droplets of such a photon condensate?  In 1920 already Einstein[8] stressed that  "...according to our present conceptions the elementary particles are ... nothing but condensations of the electromagnetic field....".

## II. BOSE-EINSTEIN CONDENSATION OF PHOTONS

From the work of Landsberg and Dunning-Davies[9] one can take a hint that in a finite space with volume  V  there could be Bose-Einstein condensation of photons.  They have calculated the  critical temperature  $T_c$  for extreme relativistic massive particles:

$$T_c = \frac{1}{k}\left[\frac{\pi^2}{\zeta(3)}\ \frac{\hbar^3 c^3 N}{g V}\right]^{\frac{1}{3}}\ ,\qquad kT_c \gg mc^2,$$

(k Boltzmann konstant,   g degeneracy).

In this formula  $T_c$  is independent of the restmass , and one can take the limit  $m \rightarrow 0$.  Of course  N  is the average number of particles, as the thermal energy is high enough for particle annihilation and creation.

Now let us consider a finite, isotropic universe with the volume  $V = 2\pi^2 R^3$, which is quasi-stationary - i.e. where it makes sense to talk of a thermodynamical equilibrium. The cosmic scale factor $R(t)$ will then vary slowly compared with the time it takes to reach the equilibrium, and we can put $R(t) = $ const. for the statistical calculations[7]. In the approximation $T \gg T_R = \dfrac{\hbar c}{kR} = 3 \cdot 10^{-29}$ K , we obtain for the number of photons

$$N = N_o + N^* = \frac{3g\Lambda}{1-\Lambda} + 2g\left(\frac{T}{T_R}\right)^3 \sum_{j=1}^{\infty} \frac{\Lambda^j}{j^4} ,$$

and the energy

$$E = 6\,kg\,\frac{T^4}{T_R^3}\sum_{j=1}^{\infty}\frac{\Lambda^j}{j^4} + 2kT_R N = 3kTN^* + 2kT_R N,$$

where the absolute activity $\Lambda$ varies between $0 < \Lambda < 1$ , and the degeneracy $g=2$ for photons with both helicity states. There is a phase transition, and the critical temperature $T_c$ is given (as before) by

$$\left(\frac{T_c}{T_R}\right)^3 = \frac{N}{2g\,\zeta(3)}$$

The number of photons not in the ground state $N^*$ is given by $N^* = (T/T_c)^3 N$ if $T \leq T_c$ , and $N^* = N$ for $T \geq T_c$ . So we obtain Planck's law for the radiation only if there is a finite proportion of the photons in the ground state, i.e. there must be a second phase. Physically we can identify this second phase only with the matter in the universe. Using the formula[10] $R^2 = \dfrac{2}{\varkappa\varrho}$ for the radius in an Einstein universe and $\varrho = 2 \cdot 10^{-29}$ g/cm$^3$ for the matter density, we obtain $N \approx N_o = \dfrac{\varrho V c^2 R}{2\hbar c} = 16 \cdot 10^{120}$ , and $T_c = 4.5 \cdot 10^{11}$ K , which corresponds to a critical mass $m_c = kT_c / c^2 = 39$ MeV. The fact that this critical mass $m_c$ is about 1/3 of the mass of the

$\pi$ meson is very surprising. If one does not consider this as purely accidental, but takes it as a connection between the structure of the universe and the properties of elementary particles, one is tempted to speculate that massive elementary particles are essentially droplets of the photon vapour. This rises immediately the question about the "surface tension" of the droplets in the super cooled vapour, which would determine the size of the droplets. However, so far  the model seems far too simple to explain  properties like half integer spin, and charge-like quantum numbers.

It seems reasonable to try to explain the time-development of the cosmos as an adiabatic expansion in our model. For fixed N we obtain $T \approx \frac{1}{R}$ , or $T = 0.9\ 10^{29}\ T_R$. This however would mean that the critical temperature had also been higher at an earlier state of the universe. It is more tempting to assume that in fact $T_c$ is a universal constant, and there has been creation of matter. We shall not pursue this question at the moment any further.

III. CONFORMAL-INVARIANT SCATTERING AT LARGE ANGLES

Nucleon-nucleon elastic scattering at high energy and fixed c.m.s. angle $\Theta \neq O, \pi$, (i.e. s, $-t$, $-u \gg 4\,m^2$, $t/s$ fixed) is a particularly interesting process for understanding the structure of massive particles in the high energy domain. If the S-Matrix becomes conformally invariant at asymptotic energies the scattering cross-section for a reaction $A + B \rightarrow C + D$ is given by

$$\frac{d\sigma}{dt} \approx \frac{1}{s^2}\left(\frac{\bar{m}^2}{s}\right)^N F(\cos\Theta),$$

where $N = x_A + x_B + x_C + x_D$ .
(This formula was derived by the author a decade ago[2,12] before there was much experimental evidence. The theory was

subsequently extended[13] and also applied to inelastic processes[14,15]. In Reference 2 the conformal quantum numbers assumed for the $\pi^\circ$ were not supported by later experimental evidence[16]. This was remarked in Ref. 17. The application to nucleon-nucleon scattering was published Ref. 18).

The scaling factor $x_i$ follows from the quantum numbers of a particular irreducible representation of the conformal group to which each particle i belongs. The (positive energy) representations can be characterized by their lowest weights[14,19]. These are the quantum numbers of the SU(2) x SU(2) x U(1) subgroup of SU(2,2) for a certain state for which the quantum number of the representation of U(1) has a minimum.

This set of quantum numbers $(j_1, j_2, n)$ has the following physical properties. In general the positive energy unitary irreducible representation of SU(2,2) has a mass spectrum $0 < m^2 < \infty$, a spin spectrum[2] $S = j_1 + j_2, \ldots \; j_1 - j_2$, and a scaling factor

$$x = n - 1 - j_1 - j_2 .$$

This was proved for mass 0 representations and massive spin 0 representations in Ref.2 and conjectured for the proton in Ref.18. Now that it has been proved for spin representation[20], the author feels confident to publish it as a general conjecture. As a special case we have $x = 0$ for mass 0 representations, and $x = \gamma + 1$ for massive spin 0 representations ($j_1 = j_2 = 0$, Casimir operator $C_I = \gamma^2 - 4$). From the

scattering experiments $pp \rightarrow pp$, $pn \rightarrow pn$, and $p\gamma \rightarrow p\pi^\circ$
follows that $x_p = x_n = 2$, $x_\pi \approx 1$, $x_\gamma = 0$.


## IV. A NEW QUARK MODEL


In this fourth section we propose a relativistic quark
model which explains why the mesons and the baryons have the
scaling factor given above. The novel feature is that we need
helicity 0 quarks $q_0$ besides the usual neutrino-like
helicity-1/2 quarks.


For the $\pi$ the experimental scaling factor $x = 1$, which
determines the representation unambiguously, $\nu = 0$. This representation is the spin-0 component of the direct product of
two mass 0, spin 0 representations. The $\pi$ reacts under the
above experimental conditions as two massless quarks $q_0$
(or quark-antiquark, as there is no difference at this level),
which however have helicity 0. Note, that the usual quark-
antiquark system $q_{-1/2} \, \bar{q}_{+1/2}$ contains no spin 0 component,
and is ruled out for this reason, too.


For the nucleon, experiments indicate $x = 2$. The product
of two helicity 0 quarks $q_0$ and a neutrino-like quark $q_{-1/2}$
leads to the following spin 1/2 representations: The symmetric
product $(q_0 q_0)$ contains the pion representation $(j_1 = j_2 = 0$,
$n = 2)$. The symmetric product $(q_0 q_0) q_{-1/2}$ of two helicity 0
quarks $q_0$, and a neutrino like quark $q_{-1/2}$ $(j_1 = 0$, $j_2 = 1/2$,
$n = 3/2)$ contains a single spin 1/2 representation $(j_1 = 0$,
$j_2 = 1/2$, $n = 7/2)$. The antisymmetric product $[q_0 q_0]$ contains

a spin 1 representation  $j_1 = 1/2$, $j_2 = 1/2$, $n = 3$. Multiplied with the neutrino $q_{-1/2}$, we obtain a representation $\left[q_o q_o\right] q_{-1/2}$ (with a spin spectrum 3/2, 1/2), characterized by the quantum numbers $j_1 = 1/2$, $j_2 = 3/2$, $n = 9/2$. This state could accomodate the nucleon N and a spin 3/2 nucleon resonance.

As both representations lead to the correct experimental scaling factor $x = 2$, we are not able to make a final decision at the moment. However, we favour the first possibility of accomodating the N in one single multiplet of this relativistic quark model. The spin 0 quarks $q_o$ obey now Bose-statistics, and one can construct a symmetric $(q_o q_o\, q_{-1/2})$ function without violating the connection between spin and statistics. In Ref.6  a model has been presented in which mass 0 particles are symmetric bound states of the lowest (conformal) energy states of the neutrino and antineutrino (urs). In this model however one cannot construct the lowest state of the helicity 0 particle. This may be the reason why the $q_o$ quarks do not occur as free physical particles.

## V. INELASTIC CHANNELS

For two incoming mass 0 particles with helicity $\lambda_1$ , and $\lambda_2$ for the elastic scattering from conformal invariance we get helicity conservation[15,21)

$$\lambda_1 + \lambda_2 = \lambda_3 + \lambda_4 ,$$

with one exception $\lambda_1 = \lambda_2 = -\lambda_3 = -\lambda_4$ , which only conserves helicity[18] mod (2). For possible additional inelasticities[14,15] we have the following results. Inelastic reactions occur only if $\lambda_1$ and $\lambda_2$ have the same sign. The total number Q of mass 0 particles in the final state is determined by the inequality

$$2 \le Q \le 2 + |\lambda_1| + |\lambda_2| - |\lambda_1 - \lambda_2| = 2 + 2\,\lambda_{min}\,\Theta(\lambda_1 \cdot \lambda_2).$$

This shows that the helicity is an essential variable at very high energies.

If the initial state consists of one mass 0, spin 0, and a massive scalar particle characterized by $\vee$ , the Casimir operator $C_I$ for the direct product has been calculated[2]. We obtain the direct sum of the representations $(j_1, j_2, n)$ :

$$(0, 0, \vee + 3) \ , \ (1/2, 1/2, \vee + 4), (1, 1, \vee + 5) \quad \text{etc.}$$

The maximal number of mass 0 particles in the final state is given by $\vee + 3$. For $\pi\gamma$ scattering we therefore expect no inelastic channels. This is a definite prediction for high momentum transfer $\pi\gamma$ scattering, and provides a check within which limits conformal invariance is a good symmetry.

Acknowledgement:

The author would like to thank P. Jacob for many discussions.

REFERENCES

1. "Deep Scattering and Hadronic Structure", ed.J.Tran Thanh Van, 1977

2. L. Castell, Phys. Rev. $\underline{D2}$ , 1161 (1970)

3. D. Sivers, S.J. Brodsky, R. Blankenbecler, Physics Reports $\underline{23C}$, 1976

4. P.A.M.Dirac, Proc.R.Soc.Lond. $\underline{A338}$, 439 (1974)

5. I. Segal: "Mathematical Cosmology and Extragalactic Astronomy", Academic Press, New York, 1976

6. L. Castell in "Quantum Theory and the Structures of Time and Space", p.147, 1975; Vol.2, p.130, 1977

7. J. Becker and L. Castell: "Photon Condensation in the Einstein Universe", Acta Physica Austriaca Suppl.XVIII, 885 (1977)

   L. Castell, in "Quantum Theory and the Structures of Time and Space 3", ed. L.Castell, C.F.v.Weizsäcker, Hanser, München 1979

8. A. Einstein, "Sidelights on Relativity", (1922),p.22

9. P.T. Landsberg and J.Dunning-Davies, Phys.Rev. $\underline{138}$, A 1049 (1965)

10. R.U. Sexl, H.K. Urbantke, "Gravitation und Kosmologie" B.J.Wissenschaftsverlag, Zürich 1975

11. S. Weinberg, "Gravitation and Cosmology", J.Wiley & Sons, New York 1972, p.619

12. L. Castell, Comm.Math.Phys.$\underline{17}$ , 127 (1970)

13. L. Castell, K.Ringhofer, in Lectures in Theoretical Physics, Vol.13, "De Sitter and Conformal Groups and Their Applications", Boulder, Colorado Assoc. University Press, 1971, p.268

14. L. Castell, in Lectures in Theoretical Physics, Vol.13 "De Sitter and Conformal Groups and Their Applications", Boulder, Colorado Assoc.University Press, 1971, p.281

15. L. Castell, Proc. X Internat.Universitätswochen für Kernphysik, Schladming, Springer, 1971, p.384

16. S.J. Brodsky, G.R. Farrar, Phys. Rev.Lett.$\underline{31}$,1153 (1973)

17. L. Castell, Nuovo Cimento Letters, $\underline{12}$, 410 (1975); $\underline{13}$, 168 (1975)

18. L. Castell, Nuovo Cimento Letters, $\underline{6}$, 115 (1973)

19.   T. Yao, J.Math.Phys.$\underline{8}$, 1931 (1967); $\underline{9}$, 1615 (1968);
      $\underline{12}$, 315 (1971)

      G. Mack, Comm.Math.Phys. $\underline{55}$, 1  (1977)

20.   P. Jacob, PH.D.Thesis, Munich 1979

21.   F. Chan, H.F. Jones, Phys.Rev. $\underline{D10}$, 1321 (1974)

# QUANTUM INTERFERENCE
# AND THE GRAVITATIONAL FIELD

Jeeva S. Anandan

Center for Theoretical Physics
Department of Physics and Astronomy
University of Maryland
College Park, Maryland

## I. INTRODUCTION

After the creation of special relativity, perhaps the most important single idea, that led to general relativity, was the equivalence of inertial and passive gravitational masses. Einstein pointed out that, in Newtonian physics, there were two apparently unrelated experimental procedures for determining the same mass. This observation ultimately led to the description of gravity by the geometry of curved space-time, which abolished the distinction between these two procedures. But curiously enough, in general relativity, there are two apparently unrelated operational procedures for determining the geometry of space-time: (i) Obtain the geometry of space-time from the behavior of clocks. This method has been emphasized by Synge[1] and Penrose[2]. (ii) Obtain the geometry from the observed classical trajectories of freely falling particles. This procedure is due to Weyl[3], Marzke and Wheeler[4] and Ehlers, Pirani and Schild[5].

As for method (i), it was pointed out by Penrose[2] that the existence of accurate clocks is ultimately due to the fact that each particle of mass m has, associated with it, a natural

157

Copyright © 1980 by Academic Press, Inc.
All rights of reproduction in any form reserved.
ISBN 0-12-473260-7

frequency $\nu_0$ given by the Einstein-Planck law

$$E_0 = mc^2 = h\nu_0 \tag{1}$$

where c is the velocity of light and h is Planck's constant.
But $\nu_0$ is determined by the behavior of the phase of the
De Broglie wave and hence depends on the propagation of this
wave. As for method (ii) a classical trajectory of a particle
is a classical limit of its quantum mechanical motion which is
described by the De Broglie wave. So if we require that the
equivalence between the two geometries is due to a common
origin, then we must probably conclude that particles have
wave nature. In other words, from an operational point of
view, gravity appears to be deeply rooted in the wave-particle
duality of matter.

Now the motion of a De Broglie wave may be regarded as
due to the _interference_ of the secondary wavelets used, via
Huygens' principle, to construct the wave. Therefore, the
gravitational field must be ultimately characterized by its
effect on the interference of a particle with itself on space-
time. This then leads us to investigate the interference of
two coherent beams in a gravitational field. We do this in
section II, in which we obtain two types of phase shifts $\Delta\phi$
and $\Delta\theta$; $\Delta\phi$ being due to the coupling of energy-momentum to the
geometry and $\Delta\theta$ due to the coupling of intrinsic spin to
space-time curvature. It is then possible to use these techniques to obtain the phase shift due to an arbitrary gauge field.
In section III, we obtain the known classical equations of
motion for a particle moving in gravitational and gauge fields,
directly from the phase shift.

In section IV, by using the experimentally known equivalence between inertial and active gravitational masses, we
argue again heuristically that there must exist a deep

connection between quantum theory and general relativity. This argument, together with the phase shift obtained in section II will suggest a modification of Einstein's field equations. Moreover it will provide an intuitive "explanation" for why the zero point energy of vacuum does not gravitate. In section V we examine whether gravity is a gauge theory in the light of the phase shift we have obtained.

## II.  PHASE SHIFT DUE TO GRAVITATIONAL AND GAUGE FIELDS

Consider a particle with spin 1/2, which will be assumed to satisfy the Dirac equation generalized to curved space-time:[6]

$$i \gamma^\mu D_\mu \psi - \frac{mc}{\hbar} \psi = 0,  \tag{2}$$

where the 4 x 4 matrix field $\gamma^\mu$ satisfy $\gamma^\mu \gamma^\nu + \gamma^\nu \gamma^\mu = 2g^{\mu\nu} I$ and $D_\mu \psi = \nabla_\mu \psi + i \Gamma_\mu \psi$ with $\nabla_\mu$ being the usual co-variant derivative on tensors and $\Gamma_\mu$ is the spin connection. To write an explicit expression for $\Gamma_\mu$, introduce a set of four vector fields $e_a{}^\mu$ (a=0,1,2,3), which are orthonormal at each point with respect to the metric with Minkowskian signature, i.e.

$$g_{\mu\nu} e_a{}^\mu e_b{}^\nu = \eta_{ab} = \mathrm{diag}(1, -1, -1, -1)$$

which is equivalent to $\eta^{ab} e_a{}^\mu e_b{}^\nu = g^{\mu\nu}$ where $g^{\mu\nu}$ and $\eta^{ab}$ are inverses of $g_{\mu\nu}$ and $\eta_{ab}$ respectively. The vierbein indices a,b... are raised and lowered using $\eta^{ab}$ and $\eta_{ab}$. Define the Ricci rotation coefficients $A_\mu{}^a{}_b = e_\nu{}^a \nabla_\mu e_b{}^\nu$. We shall assume metric compatibility, i.e. $\nabla_\mu g_{\nu\rho} = 0$, which implies that $A_\mu{}^{ab} = -A_\mu{}^{ba}$. Define also $M^{ab} = -\frac{i}{2} \left[\gamma^a, \gamma^b\right]$. Then

$$\Gamma_\mu = \frac{1}{2} A_\mu{}^{ab} M_{ab}.  \tag{3}$$

Finally, define the conjugate spinor $\bar{\psi} = \psi^+ \beta$, where $\beta = \gamma^0 = e^0{}_\mu \gamma^\mu$.

We now perform the W.K.B. approximation on (2) by requiring that $\psi = \alpha \exp(i\phi)$ where $\alpha$ and

$$k_\mu \equiv - \partial_\mu \phi$$

are slowly varying functions.  This yields[1]

$$k_\mu k^\mu = \frac{m^2 c^2}{\hbar^2} \tag{4}$$

and

$$k^\mu D_\mu \alpha = - \frac{1}{2}(\nabla^\mu k_\mu)\alpha \tag{5}$$

Equation (4) is the well known eikonal equation, whereas the
less well known equation (3) says that $\alpha$ is parallel trans-
ported along rays (integral curves of $k^\mu$) except that its
magnitude is decreasing (or increasing) depending on whether
the rays are diverging (or converging).  A wave function $\alpha e^{i\phi}$
satisfying (4) and (5) will be called a "locally approximate
plane wave."

Consider now an interference experiment in which a beam of
identical particles with fairly well defined momentum  is
split into two beams by a beam splitter  and the two beams are
made to interfere, after suitable reflections.[2]  These two
beams may be represented by locally approximate plane waves
$\psi_1 = \alpha_1 \exp(i\phi_1)$ and $\psi_2 = \alpha_2 \exp(i\phi_2)$.  The "phase difference"
$\Delta\chi$ in the region of interference will be defined by
$\exp(i\Delta\chi) = \dfrac{\bar{\psi}_1 \psi_2}{|\bar{\psi}_1 \psi_2|}$ where $|\ |$ denotes the modulus of the complex
number enclosed.  Clearly $\Delta\chi$ determines the variation of $\bar{\psi}\psi$
after interference, in small enough regions.  Now $\Delta\chi = \Delta\theta + \Delta\phi$
where $\Delta\phi = \phi_2 - \phi_1$ and

$$\exp(i\Delta\theta) = \frac{\bar{\alpha}_1 \alpha_2}{|\bar{\alpha}_1 \alpha_2|} . \tag{6}$$

---

[1] For details, see for instance reference 7.

[2] An experiment of this type with neutron beams has been
performed by Colella, Overhauser and Werner.  See Ref. 8.

Clearly

$$\Delta\phi = \oint_{\gamma} k_{\mu} \, dx^{\mu} \tag{7}$$

where $\gamma$ is a closed curve formed from two curves which lie within each of the two beams.[3]  Equation (7) is the same as the phase difference for spinless particles and has been treated elsewhere.[9],[4]  Since $k^{\mu}$ satisfies (4), $p^{\mu} \equiv \frac{\hbar}{c} k^{\mu}$ has the interpretation of energy-momentum.  Therefore (7) may be regarded as <u>the phase shift due to the coupling of the energy-momentum of the particle to the geometry</u>.  The additional phase shift $\Delta\theta$ may be interpreted as <u>the phase shift due to the coupling of spin to space-time curvature</u>.

To obtain $\Delta\theta$, note first that through any point p on the beam splitter there exist two integral curves $\gamma_{s}(\tau)$, $s = 1,2$ of the two vector fields $k^{\mu}$ corresponding to the two beams.

---

[3]It may be noted that Stokes' theorem cannot be applied to (7) because $k_{\mu}$ is discontinuous at the points where the beam is being split, reflected and interferes.

[4]It should be pointed out that in the analysis in section II of reference 9, it is only necessary to assume that the wave vector k of the locally approximate plane wave is tangent to the submanifold $\sigma$; one need not and should not assume that the waves are localized along $\sigma$ which, of course, would violate the uncertainty principle.  Also what happens during reflection is not clearly stated in this paper.  If we assume that energy is not dissipated during reflection and if $t^{\mu}$ is the four-velocity of the mirror then $k_{\mu} t^{\mu}$ should be the same for the incident and reflected wave at the point of incidence. It is then immediately clear that the entire analysis of reference 9 is valid in this case.

 

Then, from (5), $\alpha_s(x)$ for the two beams are given by

$$\alpha_s(x) = \exp(-\frac{1}{2} \int_{\gamma_s}^{x}{}_\rho \nabla^\nu k_\nu \, d\tau) \, 0\{\exp(-i \int_{\gamma_s}^{x}{}_\rho \Gamma_\mu(\tau) \frac{dx^\mu}{d\tau} \, d\tau)\} \alpha(\rho)$$

$$(s = 1,2), \qquad (8)$$

where 0 denotes path ordering which means that in the expansion of the second exponential in (8), matrices with greater value of $\tau$ occur to the left of those with smaller value of $\tau$.

If $\gamma_1$ and $\gamma_2$ form a closed curve $\gamma$ then from (6) and (8),

$$\exp(i\Delta\theta) = \frac{\bar{\alpha}_1 T \alpha_1}{|\bar{\alpha}_1 T \alpha_1|} \text{ where}$$

$$T = 0\{\exp(i \oint_\gamma \Gamma_\mu(\tau) \frac{dx^\mu}{d\tau} \, d\tau)\}.$$

It is easy now to show that when $\gamma$ is an infinitesimal curve (dropping the subscript in $\alpha$),

$$\Delta\theta = \frac{1}{2} \frac{\bar{\alpha} \phi_{\mu\nu} \alpha}{\bar{\alpha}\alpha} \, d\sigma^{\mu\nu}$$

where

$$\phi_{\mu\nu} = \partial_\nu \Gamma_\mu - \partial_\mu \Gamma_\nu + i\left[\Gamma_\nu, \Gamma_\mu\right] \qquad (9)$$

and $d\sigma^{\mu\nu}$ represents the infinitesimal area enclosed by $\gamma$. Using (3), we can also prove that

$$\Delta\theta = \frac{c}{4\hbar} S_{ab} R_{\mu\nu}{}^{ab} \, d\sigma^{\mu\nu} \qquad (10)$$

where $S^{ab} = \frac{\hbar}{c} \frac{\bar{\alpha} M^{ab} \alpha}{\bar{\alpha}\alpha}$, $R_{\mu\nu}{}^{ab} = e_\rho{}^a e_\sigma{}^b R_{\mu\nu}{}^{\rho\sigma}$, $R_{\mu\nu\rho}{}^\sigma$ being the Riemann tensor. It is now straightforward to generalize the above procedure for a particle with arbitrary spin if it obeys the Bargmann-Wigner equation [10] generalized to curved space-time. [7] In this case, $S_{ab}$ in (10) may be regarded as arbitrary. Also this phase shift (10) is in agreement with what was previously obtained using intuitive arguments. [9]

It is straightforward now to carry out this analysis for an arbritrary gauge field. [11] In this case $D_\mu \psi = \frac{\partial\psi}{\partial x^\mu} + i\Gamma_\mu \psi$ with $\Gamma_\mu = A_\mu{}^j T_j$ where $iT_j$ provide a representation for the Lie algebra of the gauge group and $A_\mu{}^j$ are the gauge potentials. A phase shift $\Delta\theta$ is obtained in the same way as before and is infinitesimally given by

$$\Delta\theta = \frac{c}{2\hbar}\, T_j\, F^j{}_{\mu\nu}\, d\sigma^{\mu\nu} \tag{11}$$

where $\tau_j = \dfrac{\hbar}{c}\, \dfrac{\alpha^+ T_j \alpha}{|\alpha^+\alpha|}$ and $F^j{}_{\mu\nu} = \partial_\nu A_\mu{}^j - \partial_\mu A_\nu{}^j - C_{\ell m}{}^j A_\mu{}^\ell A_\nu{}^m$, the $C_{\ell m}{}^j$ defined by $\left[T_\ell,\, T_m\right] = i\, C_{\ell m}{}^j\, T_j$. A special case of this phase shift is the Aharonov-Bohm effect.[12] As in the original Aharonov-Bohm effect, the phase shift can be non zero, during interference in a multiply connected region, for all $\Delta\theta$ type of phase shifts, including the above phase shift due to the spin-curvature coupling.

We shall also show now, using intuitive arguments, that in a space-time that has non-zero torsion, there is a phase shift, in addition to (10), which is due to the coupling of energy-momentum to torsion. Notice first that the identity $2\, Q_{\nu\rho}{}^\mu x^\nu y^\rho = x^\nu \nabla_\nu y^\mu - y^\nu \nabla_\nu x^\mu - [x,y]^\mu$, where $Q_{\nu\rho}{}^\mu = \Gamma^\mu{}_{[\nu\rho]}$ is the torsion, implies that if x and y are infinitesimal vectors and we try to construct an infinitesimal parallelogram by parallel transporting x along y and y along x then the "parallelogram" will fail to close by an amount $\Delta x^\mu = Q_{\nu\rho}{}^\mu\, d\sigma^{\nu\rho}$ where $d\sigma^{\nu\rho} = 2\, x^{[\nu} y^{\rho]}$. Intuitively, it is then clear that torsion will modify (7) by an amount which is infinitesimally given by

$$\Delta\theta = k_\mu\, Q_{\nu\rho}{}^\mu\, d\sigma^{\nu\rho} = \frac{c}{\hbar}\, P_\mu\, Q_{\nu\rho}{}^\mu\, d\sigma^{\nu\rho} \ . \tag{12}$$

A mathematically more precise argument leading to (12), which also shows the relationship between (10) and (12) and the Poincare group, will be given in section V.


III.  CLASSICAL LIMIT


We shall now demonstrate the usefulness of the above phase shifts by obtaining all the known basic equations of motion for classical particles in external fields, directly from the phase shift. To do so, consider the "correspondence principle" between the phase shift and the classical equation of motion given previously.[9,13] A covariant generalization of this relation is

$$\Delta\theta \, v^{\mu} = \frac{c}{\hbar} \, d\sigma^{\mu\nu} \, \frac{DP_{\nu}}{Ds} \tag{13}$$

where $v^{\mu}$ is the four-velocity of the classical trajectory, P is the energy-momentum $\frac{D}{Ds}$ denotes covariant differentiation with respect to proper time s and $\Delta\theta$ is the phase shift due to the external field between interfering beams spanning the infinitesimal area $d\sigma^{\mu\nu}$ which must contain $v^{\mu}$. The last requirement implies that

$$d\sigma^{\mu\nu} = u^{[\mu} v^{\nu]} \tag{14}$$

for some infinitesimal vector $u^{\mu}$ and $[\quad]$ denotes anti-symmetrization. It is easy to show that in a local Minkowski coordinate system in which $v^{\mu} = \gamma(1, \frac{v}{c}, 0, 0)$, $\gamma = (1 - \frac{v^2}{c^2})^{-1/2}$, $\Delta\theta$ given by (13) can be written for a massive particle as $\Delta\theta = \Delta\theta_{\perp} + \Delta\theta_{\parallel}$ where $\Delta\theta_{\perp} = \frac{m\gamma A}{\hbar v} \frac{dv}{dt}_{\perp}$ and $\Delta\theta_{\parallel} = \frac{m\gamma^3 \Sigma}{\hbar c} \frac{dv}{dt}_{\parallel}$, $\frac{dv}{dt}_{\perp}$ and $\frac{dv}{dt}_{\parallel}$ being the components of the acceleration perpendicular to and parallel to the velocity, A and $\Sigma$ being projections of $d\sigma^{\mu\nu}$ on appropriate two dimensional space-like and time-like planes. The expression for $\Delta\theta_{\perp}$ is the same as (24') of reference 9, which is associated with the transverse acceleration. $\Delta\theta_{\parallel}$ is the phase shift associated with the longitudinal acceleration $\frac{dv}{dt}_{\parallel}$. For instance, the acceleration of an electric charge, moving along the x-direction, due to an electric field in the x-direction, can be related to the phase shift in interference in the x-t plane by means of the latter relation. This shows that (13) and (14) not only covariantly generalize the previously given relation for $\Delta\theta_{\perp}$ but also yield a new relation. Also these equations relate the holonomy, which determines $\Delta\theta$, to the classical equation of motion, and is therefore different from other treatments of the classical limit, such as the Hamilton-Jacobi equation.

It is now straightforward to show using (13), (14) and the arbitrariness of the direction of $u^{\mu}$ that the classical limit corresponding to (10) is

$$\left(\frac{DP^{\mu}}{Ds}\right)_{curvature} = \frac{1}{2}\,S_{ab}\,R^{\mu}{}_{\nu}{}^{ab}\,v^{\nu} = \frac{1}{2}\,S_{\rho\sigma}\,R^{\mu}{}_{\nu}{}^{\rho\sigma}\,v^{\nu} \quad (15)$$

which is the Mathisson-Papapetrou[14] equation. Since (15) is a consequence of (10) which can be deduced from the Bargmann-Wigner equation, we have shown, perhaps for the first time, that the Mathisson-Papapetrou equation is the classical limit of Bargmann-Wigner equation on curved space-time. Similarly from (11)

$$\left(\frac{DP^{\mu}}{Ds}\right)_{\substack{gauge \\ field}} = \tau_{j}\,F^{j\mu}{}_{\nu}\,v^{\nu}\ , \quad (16)$$

which was first given by Wong.[15] For a particle moving in a space-time with torsion, there is an additional phase shift (12) which in the classical limit, via (13) and (14), yields

$$\left(\frac{DP^{\mu}}{Ds}\right)_{torsion} = 2\,P_{\rho}\,Q^{\mu}{}_{\nu}{}^{\rho}\,v^{\nu}. \quad (17)$$

The total force experienced by this particle is then the sum of (15) and (17), which was first given by Hehl[16] in a slightly different form. This equation and (15) were previously obtained using conservation laws, which is different from our considerations which lead to the same equations. Also it is quite remarkable that all these equations of motion can be obtained from the same general concept, namely the phase shift in quantum interference, which incidentally also suggests their quantum mechanical origin.

## IV.  FIELD EQUATIONS

Consider now equation (16) which says that there is a local change in energy-momentum of the particle as a result of the phase shift $\Delta\theta$ experienced.  Since the total energy-momentum is conserved, there must be a local exchange of energy-momentum between the particle and the field.  Therefore, we must associate an energy-momentum tensor with the gauge field.  Also we note that the reaction of the particle back on the field, gives us the field equations.  Indeed, by considering the local exchange of energy-momentum between the particle and the gauge field it is easy to show that the simplest field equations for the gauge field are the Yang-Mills equations.[7]  We may therefore associate the usual Lagrangian for the Yang-Mills field with the $\Delta\theta$ phase shift, a remark that will become important later on.

We now generalize the above argument to the following heuristic principle:  <u>A particle acts on a field (field equations) in a manner that depends on its response to the field (equation of motion or the phase shift)</u>.  In general relativity, however, the trajectory of a freely falling massive particle is independent of its mass and yet the mass modifies the geometry and hence the gravitational field.  However, we can reconcile the latter fact with the above principle if we take the motion of the particle to be not merely the unparametrised world-line but rather a parametrised world-line with the parameter depending on the mass.  The particle then carries with it some sort of "natural clock" such that its frequency $\nu_g$, or the number of "ticks" with respect to some standard clock, depends on the particle's mass.  The simplest assumption now is to suppose that the energy $m_g c^2$ of the particle in its

rest frame is proportional to $\nu_g$, since we know that more

generally, for both massive and massless particle, the energy-

momentum acts as a source of gravity. The constant of propor-

tionality can now be made equal to Planck's constant by appro-

priately choosing the standard clock in the definition of $\nu_g$

and we obtain the equation

$$m_g c^2 = h\nu_g.$$

The mass $m_g$ in the above argument is the active gravitational

mass since it is the source of gravity. But if we now iden-

tify $\nu_g$ with the frequency $\nu_o$ associated with the phase of the

wave function of the particle in its rest frame, then $m_g$ must

also be equal to the inertial mass m and vice versa.[5] This

is because, it is clear from the treatment in the last section,

that the acceleration of the particle in an external field is

inversely proportional to the mass m in equation (1) so that

the latter mass is the inertial mass of the particle. Hence

we obtain, from the heuristic principle stated above, the equi-

valence principle and simplicity, equation (1) which is perhaps

the most basic equation in quantum theory. Conversely, let

us assume quantum theory but pretend that we are ignorant of

gravity and general relativity. When we do an ordinary inter-

ference experiment, such as the Young's double slit

---

[5] More generally, the two frequencies $\nu_o$ and $\nu_g$ give two

time scales that are associated with the phase of the wave

function and the above "natural clock", which may respectively

be called atomic and gravitational times, such that the ratio

of these two times is equal to the ratio of inertial and

active gravitational masses $m_o$ and $m_g$. So if one of them is not

a constant then the same is true of the other.

experiment, we find that there is a phase difference $\Delta\phi$ which depends on the energy-momentum and the distances traveled by the beams, i.e. on the geometry. Equation (7) is just a covariant expression of this well known phase difference. Then the above principle implies that the interfering particle should modify the geometry in a manner that depends on its energy-momentum. This leads to the basic idea in general relativity and implies that the particle must act as a source of gravity. Hence general relativity and quantum theory are by themselves incomplete and need each other, if the above principle is to be satisfied. Since according to the first part of the above argument the matter field must be treated quantum mechanically, its reaction back on the gravitational field in accordance with the above principle would require that the latter field also be treated quantum mechanically.

The above argument, according to which gravity is deeply rooted in the wave particle duality of matter, has as a consequence the equivalence of inertial and active gravitational mass. This is because the motion of the wave associated with the particle depends on its inertial mass. Its reaction back on the gravitational field must then also be proportional to the same mass. Also this provides an intuitive understanding of why the zero point energy of vacuum does not gravitate. This is because particles that have wave functions, which interfere and therefore gravitate correspond to excitations above the zero point energy.

To be sure, the above arguments are of an intuitive and heuristic nature. Nevertheless, they seem to suggest strongly that the gravitational field equations are quantum mechanical in origin and that the phase shift in quantum interference may

be a key to understanding the field equations.  Now $\Delta\phi$ itself,
in the classical limit, would give the geodesic equation,[6] so
that there is no local exchange of energy-momentum between the
particle and the field; therefore the energy-momentum of the
matter field is locally conserved.  We can then follow the
usual procedure[17] and obtain Einstein's field equations as
the simplest field equations according to which, as we have
already concluded, matter must influence geometry.   We may
therefore associate the usual Ricci scalar Lagrangian with $\Delta\phi$.
But the additional phase shift $\Delta\theta$ due to the coupling of spin
to curvature, however, yielded (15).  This derivation of (15)
implies that it is valid for a point particle with <u>intrinsic</u>
<u>spin</u> so that there must be a local exhcange of energy-momentum
between the particle and the gravitational field.  We saw
earlier that a similar argument for gauge fields led to
associating with the $\Delta\theta$ type of phase shift a Lagrangian
that is quadratic in the field strength, on grounds of
simplicity, which has also been amply confirmed by experiment
for the electromagnetic case.  Then the principle stated in
the second paragraph of this section leads us to describe the
reaction of the particle to the $\Delta\theta$ phase shift in gravity also
by a similar term in the Lagrangian which is quadratic in the
curvature.  Also if we allow spin to react back on the curvature in a manner which is on the same dynamical footing as the

---

[6]This is clear if we go to the local inertial frame where
$\Delta\phi$ is infinitesimally like the phase difference between interfering beams in flat space-time, so that the classical limit
is a curve which in this frame is locally a straight line, i.e.
a geodesic.

reaction of the energy momentum to the geometry then the
connection and metric must be independent variables.  i.e.,
we must introduce torsion.  These arguments lead us to
regard

$$L = \frac{1}{16\pi K} \, R_{\alpha\beta}{}^{\beta\alpha} + \frac{1}{32\pi f} \, R_{\alpha\beta\mu}{}^{\nu} \, R^{\alpha\beta}{}_{\nu}{}^{\mu} + L_M \tag{17}$$

where K, f are constants, $L_M$ is the matter Lagrangian and the
Riemann tensor contains torsion, as the simplest Lagrangian
for the gravitational field consistent with our general principles.  The first two terms in (17) are in our view, associa-
ted with $\Delta\phi$ and $\Delta\theta$ respectively.  By varying the metric and
the connection in (17) independently, but assuming metric compatibility, one obtains the following two field equations with
the energy-momentum and spin densities respectively as the
sources:

$$G^{\mu\nu} = 8\pi K (T^{\mu\nu} - \frac{1}{8\pi f} \, R^{\mu}{}_{\rho\alpha}{}^{\beta} \, R^{\nu\rho}{}_{\beta}{}^{\alpha}$$

$$+ \frac{1}{32\pi f} \, g^{\mu\nu} \, R^{\alpha\beta}{}_{\gamma}{}^{\delta} \, R_{\alpha\beta\delta}{}^{\gamma}) \tag{18}$$

and

$$\nabla_{\tilde{\nu}} \, R^{\tilde{\mu}\tilde{\nu}}{}_{\alpha\beta} = 4\pi f \, S_{\alpha\beta}{}^{\tilde{\mu}} - \frac{f}{K} \, (Q_{\alpha\beta}{}^{\tilde{\mu}} + 2\delta_{[\alpha}{}^{\tilde{\mu}} Q_{\beta]}{}_{\sigma}{}^{\sigma}) \tag{19}$$

where the tilde over an index means that during parallel
transport this index must be corrected by the Christoeffel
connection and $S^{\alpha\beta\hat{\mu}}$ is the spin density.  A theory of gravity
in which torsion is algebraically coupled to the spin density
has been studied by Sciama and Kibble[18].  In the present
theory, however, torsion is differentially coupled to spin
density and consequently it propagates.

Theories of the type (17) have been considered by
Mansouri and Chang[19] based on the elegant notion of fibre
bundle geometry.  The spinor form of this theory has also
been considered by Fairchild[20], who was the first to
realize the need to introduce torsion into such a theory.

Fairchild introduced the second term in (17) to obtain a Yang-Mills type of theory and the first term to have the Newtonian limit.[7] In our approach[8], however, both terms arise from the same general consideration and an appeal to Newtonian limit is not necessary. Ramaswamy and Yasskin[21] have proved an analogue of Birkhoff's theorem for this theory: for an 0(3) spherically symmetric solution of this theory, torsion must be zero and the metric is Schwarzschild. These results suggest that this theory satisfies all the experimental tests satisfied at present by Einstein's theory; but as

---

[7] It has been mentioned in reference 21 that Fairchild has not shown the Newtonian limit because he does not justify the identification $\rho = T_{oo} + V_{oo}$ where $V_{oo}$ represents torsion, as the mass density whereas it is $T_{oo}$ that is normally considered to be the mass density. However, to have the Newtonian limit it is necessary and sufficient that, in the weak field limit, the gravitational field due to a particle is inversely proportional to the square of the distance from the particle and the constant of proportionality divided by K is, by definition, the active gravitational mass. This leaves unanswered the question of whether the active gravitational mass is proportional to the inertial mass in this theory within the frame-work considered in references 20 and 21. From our point of view, however, since gravity is due to the wave nature of particles, there must exist particles associated with torsion, corresponding to $V_{oo}$. Moreover the argument in Section IV that led to the equality of inertial and active gravitational masses answers the above question in the affirmative.

[8] See reference 7 for a more detailed account of this approach.

we saw in the present paper the former is more natural than
the latter when one considers  the quantum mechanical interaction between a particle and the gravitational field.

V.   IS GRAVITY A GAUGE THEORY?

We ask now, whether the phase shifts $\Delta\phi$ and $\Delta\theta$ for the
gravitational field described above can be treated in a
unified manner.  We noted earlier that because of (4),
$P_\mu \equiv \frac{\hbar}{c} k_\mu$  may be regarded as the eigen values of the energy-
momentum operator.  This suggests a description of gravity
in terms of the integral formalism of gauge fields.[22]  In
this formalism, a gauge field is described by associating with
each path in space-time an element of the corresponding
gauge group.  For the gravitational field we may associate
with each path C on space-time an element of the covering group
of the Poincare group (generated by SL(2, C) and translations)

$$\Lambda_C = O\{\exp(-i \int_C \bar{\Gamma}_\mu \, dx^\mu)\} \tag{20}$$

with $\bar{\Gamma}_\mu = e_\mu{}^a P_a + \frac{1}{2} A_\mu{}^{ab} M_{ab}$ where $ip_a$, $iM_{ab}$ generate the Lie
algebra of the Poincare group:

$$[M_{ab}, M_{cd}] = i(\eta_{bc} M_{ad} - \eta_{ac} M_{bd} + \eta_{ad} M_{bc} - \eta_{bd} M_{ac}),$$

$$[P_a, P_b] = 0, \quad [M_{ab}, P_c] = i(\eta_{bc} P_a - \eta_{ac} P_b).$$

During an interference experiment, the evolution of a locally
approximate plane wave, between reflections, is determined by
the corresponding representation of an operator of the form
(20).  This provides a unified way of obtaining the gravitational phase shift.  It may appear that gravity is not a gauge
field in the differential formalism of gauge fields, because
gravity is described not only by a connection but also a
metric.  However, following Kibble[18,23] we may write
$e_\mu{}^a = \delta_\mu{}^a + B_\mu{}^a$ where $B_\mu{}^a$ may be regarded as the gauge
potential of the translation group.

But the unsatisfactory feature in this scheme is that the
above decomposition of $e_\mu{}^a$ is coordinate dependent.  So the
integral formalism is superior to the differential formalism
because this decomposition is not necessary in (20).  From
this point of view, gravity may be regarded as the manifesta-
tion of a non integrable phase factor of the covering  group
of the Poincare group.[9]  In describing gravity this way, however, we have generalized the usual integral formalism, since
previously the gauge groups used in this formalism were
"internal" groups which leave space-time points invariant,
unlike the poincare group.  In this respect gravity differs
from other gauge fields.  Nevertheless, the present approach
provides a unified description of gravity and gauge fields by
associating an appropriate element of the entire invariance
group with each path.

We note finally that if C is an infinitesimal closed curve,
spanning $d\sigma^{\mu\nu}$, (20) gives, on using the Poincare algebra,

$$\Lambda_c = 1 + \frac{i}{2}(2\,Q_{\mu\nu}{}^a\,P_a + \frac{1}{2}\,R_{\mu\nu}{}^{ab}\,M_{ab})\,d\sigma^{\mu\nu} \tag{21}$$

where $Q_{\mu\nu}{}^a \equiv e_{[\mu,\nu]}{}^a - \frac{1}{2}\,A_\mu{}^a{}_c\,e_\nu{}^c + \frac{1}{2}\,A_\nu{}^a{}_c\,e_\mu{}^c = e_\rho{}^a\,Q_{\mu\nu}{}^\rho$
is the torsion form.  In differential geometric terms, we
have obtained here the known result that the development of
an infinitesimal closed curve in the <u>affine</u> bundle is  determined by the curvature and torsion.[24]  This result again

yields the phase shifts (10) and (12) and show their connection with the Poincare group.  But it should be emphasized

---

[9]The existence of fermions requires that (20) is an element of
the covering group of the Poincare group.  This can also be
demonstrated by an interference experiment, as done by Werner,
Colella, Overhauser and Eagen, Phys. Rev. Lett. <u>35</u>, 1053(1975).

that it is metric compatibility which restricts the holonomy
group to a subgroup of the Poincare group.  Otherwise the
holonomy group could be any subgroup of the affine group
(generated by GL(4) and the translations) which must then be
regarded as the gauge group.  An important advantage of the
present approach, however, is that metric compatibility is a
natural consequence of our view that the metric is quantum
mechanical in origin as opposed to classical general relativity where metric compatibility must be postulated.  This
is basically because, according to the discussion in sections
I and IV, the metric is determined from the mass.  And the
square of the mass is $\eta^{ab} P_a P_b$ which is a Casimir operator
of the Poincare group and is consequently invariant under
(20).  This is also, in our view, the basic reason why mass
is a good quantum number in the presence of gravity.  Thus
even though classically, gravity breaks Poincare invariance,
so that the Poincare group then becomes irrelevant, quantum
mechanically gravity should be described by the (covering
group of the) Poincare group at least in a limiting sense.
The details of some of the above arguments will be given in a
future paper.


ACKNOWLEDGMENTS

I wish to thank C. W. Misner, J. M. Nester and P. B.
Yasskin for clarifying discussions.

REFERENCES


1.  J. L. Synge, <u>Relativity: The General Theory</u> (North Holland, Amsterdam, 1960), Chapter III.

2.  R. Penrose, In <u>Battelle Recontres</u>, (ed. C. M. DeWitt and J. A. Wheeler; Benjamin, New York 1968), section 2.

3.  H. Weyl, Nachr. Ges. Wiss. Gottingen, <u>99</u>, (1921)

4.  R. F. Marzke and J. A. Wheeler, in <u>Gravitation and Relativity</u> (ed. H. Y. Chiu and W. F. Hoffmann: Benjamin, New York, 1964).

5.  J. Ehlers, F.A.E. Pirani and A. Schild in <u>General Relativity</u> (ed. L. O'Raifeartaigh, Clarendon, Oxford 1972).

6.  P.A.M. Dirac, Proc. R. Soc. Lond. <u>A180</u>, 1, (1942); V. Fock and D. Ivanenko, Compt. Rend. <u>188</u>, 1470 (1929).

7.  J. Anandan, Univ. of Maryland Report PP#79-139 (to appear in Il Nuovo Cimento, 53A, 221 (1979).

8.  A. W. Overhauser and R. Colella, Phys. Rev. Lett. <u>33</u>, 1237 (1974); R. Colella, A. W. Overhauser and S. A. Werner, Phys. Rev. Lett. <u>34</u>, 1472 (1975).

9.  J. Anandan, Phys. Rev. D, <u>15</u>, 1448 (1977).

10.  V. Bargmann and E. P. Wigner, Proc. National Acad. Sciences, <u>34</u>, 211 (1946).

11.  See also T. T. Wu and C. N. Yang, Phys. Rev. D. <u>12</u>, 3845 (1975).

12.  Y. Aharonov and D. Bohm, Phys. Rev. <u>115</u>, 485 (1959).

13.  See also The Feynman Lectures on Physics (Addision-Wesley, 1964), Vol. II, Sec. 15-5, Vol. III, Sec. 7-4.

14.  M. Mathisson, Acta Phys. Polon. <u>6</u>, 163 (1973); A. Papapetrou, Proc. R. Soc. London <u>A209</u>, 258 (1951).

15.  S. K. Wong, Il, Nuovo Cimento LXVA, 689 (1970).

16.  F. W. Hehl, Physics Letters, 36A, 225 (1971).

17.  See for instance C. W. Misner, K. S. Thorne and
     J. A. Wheeler, Gravitation (Freeman 1973) §17.1.

18.  D. W. Sciama in Recent Developments in General Relativity
     (Pergamon, Oxford 1962); T.W.B. Kibble, J. Math. Phys. 2,
     212 (1961).

19.  F. Mansouri and L. N. Chang, Phys. Rev. D., 15, 3192
     (1976).

20.  E. E. Fairchild Jr., Phys. Rev. D., 16, 2438 (1977).

21.  S. Ramaswamy and P. B. Yasskin, Phys. Rev. D., 19,
     2265 (1979).

22.  C. N. Yang, Phys. Rev. Letters, 33, 445 (1974).  See also
     P.A.M. Dirac, Proc. of Roy. Soc. Lond. A133, 60 (1931);
     S. Mandelstam, Ann. Phys. 19, 1, 25 (1962);
     I. Bialynicki-Birula, Bull. Acad. Pol. Sci. 11, 135
     (1963).

23.  See also Y. M. Cho, Phys. Rev. D. 12, 3341 (1976).

24.  See for instance S. Kobayashi and K. Nomizu,
     Foundations of Differential Geometry, Interscience,
     New York (1963), Ch. III.

# GEOMETRIZATION OF GAUGE FIELDS

George F. Chapline[1]

Theoretical Physics Division
Lawrence Livermore Laboratory
University of California
Livermore, California

I would first like to comment on the problem of the non-renormalizability of general relativity and supergravity theories (1). It is possible that this is merely a technical problem that can be overcome in some non-perturbative formulation of quantum gravity. On the other hand, it may be an indication that the graviton is not an elementary particle; just as the non-renormalizability of hydrodynamics is a reflection of the fact that vortices are not elementary particles.

In this talk I would like to adopt the point of view that the graviton is not an elementary particle but instead is a collective excitation of the vacuum. In particular, I would like to put forward the idea that gravity is a collective excitation of non-abelian gauge fields.

That gauge fields can form geometrical objects has been known for some time. The simplest example is the flux tube in a type II superconductor (2). These flux tubes can vibrate and have a life of their own. A more sophisticated example

---

[1]Work performed under the auspices of the U. S. Department of Energy by the Lawrence Livermore Laboratory under contract number W-7405-ENG-48.

Copyright © 1980 by Academic Press, Inc.
All rights of reproduction in any form reserved.
ISBN 0-12-473260-7

is the t'Hooft-Polyakov monopole (3) in gauge theories with
a spontaneously broken SU(2) gauge symmetry.

The flux tube in a superconductor and the t'Hooft-Polyakov
monopole delineate regions of "normal" vacuum imbedded inside
regions with a superconducting vacuum.  The superconducting
vacuum differs from the normal vacuum in that a gauge transformation acting on a superconducting vacuum leads to a
different degenerate vacuum state, whereas the normal vacuum
is invariant under gauge transformations.  The form of the
boundary between normal vacuum regions and superconducting
vacuum regions is related to the topological properties of
the manifold of degenerate vacuum states (4).  In particular,
flux lines can form when the lowest non-trival homotopy group
for this manifold is the fundamental group $\pi_1$.  In the case of
spontaneously broken SU(2) the lowest non-trival homotopy
group is $\pi_2$ so the normal vacuum is confined to a point.

In general if a gauge group G is spontaneously broken the
form of the boundary between normal and superconducting
regions will be determined by the lowest non-trival homotopy
group $\pi_k$ for the coset space G/H, where H is the little group
under which superconducting vacuum states are invariant.  It
is not hard to show that in n+1 dimensional space-time the
normal vacuum will be confined to a region of spacelike
dimensionality n-k-1.

Of primary interest to us here is the question of whether
a normal vacuum with three space-like dimensions can arise
naturally and whether such a normal vacuum can be regarded
as a 4-dimensional Riemanian manifold obeying Einstein's
equations.  In response to the first of these questions we
can offer some examples of spontaneous symmetry breaking

that might be expected to lead to 3+1 dimensional regions with normal vacuums.  Two such examples are (a) spontaneous breaking of an O(6) to O(5) gauge symmetry in 10-dimensional spacetime, and (b) spontaneous breaking of an O(10) to SU(5) gauge symmetry in 26-dimensional spacetime.  There is an amusing similarity between the first example and N=4 supergravity; which can be obtained from simple supergravity in 10-dimensional spacetime by "dimensional reduction" and posses an O(6) internal symmetry (5).  The second example is interesting because both O(10) and SU(5) have been discussed as possible candidates for a unified field theory of strong, weak, and electromagnetic interactions (6).  In addition general relativity is string renormalizable in 26-dimensional spacetime.

In regard to the second question we can only offer a plausibility argument at the present time.  It should first of all be pointed out that even if the normal vacuum is confined to a region with 3 spacelike dimensions it does not immediately follow that the normal vacuum satisfies Einstein's equations.  For example, the relativistic motion of a flux line (7) is described by the Nambu Lagrangian - $\int \sqrt{-g}\ d^2\sigma$, not 2 dimensional gravity.

It is possible, however, to introduce a non-trivial dependence of the action for a string on curvature by introducing a 4-dimensional spinor field on the string, together with a spin-spin interaction (8).  We might expect that a similar situation would prevail for 4-dimensional continua with non-abelian gauge fields.  In particular, a vector gauge field could act as a higher dimensional spin field which sensed the extrinsic curvature of a 4-dimensional continuum.

If the vector field is non-abelian then a "spin-spin" interaction would arise if the nonlinear terms in the Yang-Mills
Lagrangian are non-zero.  We prove in an Appendix that nonlinear terms in the Lagrangian will in general always occur
when normal vacuum regions are embedded in superconducting
vacuums.

Thus if we have a 4-dimensional normal vacuum imbedded
in a higher dimensional flat spacetime, one could measure the
intrinsic curvature of this vacuum via the Gauss-Codazzi
equations by measuring the changes in the vector gauge fields
as one moves through the vacuum.  The nonlinear terms in the
Yang-Mills Lagrangian in the case of non-abelian vector gauge
fields would then introduce a dependence of the action on the
intrinsic curvature of spacetime.

In summary then, we propose to model our own universe by
looking at solutions of the Yang-Mills equations in cases
where the normal vacuum solutions are confined to a 4-dimen-
sional, possibly curved, subspace of a higher dimensional
flat spacetime.  As a specific example, one might consider
spontaneously broken O(6) gauge symmetry in 10-dimensional
spacetime.  The Lagrangian for this system is

$$L = -\tfrac{1}{2}(D_\mu\phi)^2 - \frac{\lambda}{4}(\phi^2 - \phi_0^2)^2 - \tfrac{1}{4}F^a_{\mu\nu}F^{\mu\nu}_a$$

where $D_\mu\phi^a \equiv \partial_\mu\phi^a + ef_{abc}A^b_\mu\phi^c$, $F^a_{\mu\nu} \equiv \partial_\mu A^a_\nu - \partial_\nu A^a_\mu + ef_{abc}A^b_\mu A^c_\nu$,
$f_{abc}$ are the O(6) structure constants, and $\phi^a$ and $A^a_\mu$ are
scalar and vector fields corresponding to the fundamental
representation of O(6).  The analogue of the superconductor
flux line for this case would be a 6-dimensional solution to
the static scalar field—Yang-Mills field equations derived
from this Lagrangian.  Normal vacuum solutions of the O(6)

Yang-Mills equations would be trapped inside the 3-dimensional
tube formed by the 6-dimensional static solution.

Apart from the aesthetically pleasing feature of being
able to explain gravity without having to introduce gravitons
as elementary particles one should, of course, ask if the
present picture makes any difference observationally.  In
fact, it appears that the present ideas may have profound
cosmological consequences.  In particular, the author has
recently suggested that the "big bang" might be interpreted
as the nucleation of a normal vacuum region in a higher
dimensional space (9).

This interpretation permits one to make rough predictions
concerning the properties of our own universe.  For example,
the initial curvature of our universe would be determined by
the average energy density in the normal vacuum region just
after it formed.  This average energy density will in turn be
determined by the critical field for the gauge theory (10).
Now it turns out that there are two kinds of critical fields
in Ginzburg-Landau-Higgs type theories, depending on whether
$e^2 > 2\lambda$ or $e^2 < 2\lambda$.  The case of interest to us is $e^2 < 2\lambda$ in
which case the critical field in the 0(6) case is approxi-
mately $e\mu^2/\lambda$.  The corresponding energy density is approximately $e^2\mu^4/2\lambda^2$.

If we identify the Planck mass with $4\mu/e^2$, which is
approximately the initial nuclear mass of a normal vacuum
region, then the energy density which would just cause the
initial normal vacuum region to be gravitationally bound is
$\approx 4\mu^4/\lambda^2$.  The curvature parameter $\Omega_o$, defined as the ratio
of the actual energy density in an initial normal vacuum

region to this "parabolic" energy density would be approximately given by:

$$\Omega_o \approx \frac{e^2}{8} \quad .$$

If we identify $e^2/4\pi$ with the coupling constant in a unified theory of weak, electromagnetic, and strong interactions (6) then $e^2/4\pi \approx .02$ and so $\Omega_o \approx .03$. This is not far from the value of the curvature parameter inferred, e.g. from deuterium abundances, for our own universe. We are led to the exciting possibility that the large scale curvature of our universe may be related to fundamental physical constants. In conclusion, associating our universe with a normal vacuum region may have interesting consequences and deserves further study.

REFERENCES

1.    Veltman, M., "Methods in Field Theory" (R. Balian and J. Zinn-Justin, eds.).  North-Holland, Amsterdam, (1976).

2.    Abrikosov, A. A., *Sov. Phys. JETP 5*, 1174 (1957).

3.    't Hooft, G., *Nucl. Phys. B79*, 276 (1974); Polyskav, A. M., *Sov. Phys. JETP Lett. 20*, 194 (1974).

4.    Kibble, T. W. B., *J. Phys. A9*, 1387 (1976).

5.    Gliozzi, F., Scherk, J., and Olive, D., *Phys. Lett. 65B*, 282 (1976).

6.    Georgi, H., and Glashow, S. L., *Phys. Rev. Lett. 32*, 438 (1974).

7.    Nielsen, H. B. and Olesen, P., *Nucl. Phys. B79*, 276 (1973).

8.    Ng, Y. J. and Tye, S. H., *Phys. Rev. D16*, 2468 (1977).

9.   Chapline, G., Lawrence Livermore Laboratory Report
     UCRL-82720, Livermore, California (1979).

10.  Krive, I., Pyzh, V. and Chudnovskii, E., *Sov. J. Nucl.
     Phys. 23*, 358 (1976).

11.  Sikivie, P., SLAC Preprint 2287, Stanford University,
     Stanford, California (1979).

APPENDIX

It has been shown by Sikivie (11) that abelian magnetic field solutions of the Yang-Mills equations in 3+1 dimensions are unstable if the magnetic field is sufficiently strong. We shall show here that an analogous result holds for normal vacuum regions embedded in a higher dimensional space. For concreteness we shall concentrate on the O(6) Yang-Mills-Higgs system in 9+1 dimensions discussed in the text.

Let us suppose that there is a static magnetic field of strength B existing throughout a particular 3-dimensional space-like subspace of the 9+1 dimensional spacetime. In order to simplify our discussion we shall assume that this subspace is just the Euclidean subspace $R^3$ spanned by the 1, 2, and 3 axes. Outside the subspace $R^3$ there will be a nonvanishing Higgs field $\phi$. Now let us assume that the magnetic field is abelian, i.e., oriented along constant direction in the O(6) internal space. We shall show that this situation is unstable in that there are non-abelian components which grow with time.

The magnetic field in the 3-dimensional region $R^3$ can be expressed in terms of an abelian vector potential $A_j$:

$$B_{ij} = \partial_i A_j - \partial_j A_i \tag{A.1}$$

where $i,j = 1,2,3$.  On the other hand, for $\lambda > e^2/2$ there will
be some leakage of the magnetic field into the 6-dimensional
space orthogonal to $R^3$ and so we may consider using Eq. (A.1)
to define $B_{ij}$ for $i,j = 1,\ldots,9$ (we will use the $A_o = 0$ gauge).
Let us now consider small oscillations $a_i^b$ around the background field $A_j$:

$$A_i^a = \delta^{al}A_j + a_j^a \tag{A.2}$$

where $i,j = 1,\ldots,9$ and $a = 1,\ldots,15$.  To first order in $a_j^a$
the electric and magnetic fields are

$$E_j^a = \dot{a}_j^a \quad,$$

$$B_{ij}^a = \delta^{al}(\partial_i A_j - \partial_j A_i) + \partial_j a_j^a - \partial_j a_i^a$$

$$- ef_{lab}(A_i a_j^b - A_j a_i^b) \quad.$$

The linearized Yang-Mills equations are

$$\partial_i \dot{a}_i^a - ef_{lab}A_j \dot{a}_i^b = 0 \tag{A.3a}$$

$$\ddot{a}_j^a - \nabla^2 a_j^a + \partial_j \partial_i a_i^a + e^2(\delta_{ab} - \delta_{al}\delta_{bl})(A_i A_i a_j^b - A_j A_i a_i^b)$$

$$+ ef_{lab}\left[(\partial_i A_i)a_j^b - (\partial_i a_i^b)A_j + 2A_i \partial_i a_j^b\right.$$

$$\left. - 2(\partial_i A_j)a_i^b + (\partial_j A_i)a_i^b - A_i \partial_j a_i^b\right]$$

$$+ e^2 \phi_o^2 f_{lab}a_j^b = 0 \quad, \tag{A.3b}$$

where $\phi_o$ is the vacuum expectation value of the Higgs  field.
Integrating Eq. (A.3a) we obtain

$$\partial_i a_i^a - ef_{lab}A_i a_i^b = 0 \quad.$$

Combining this with Eq. (A.3b) yields

$$\ddot{a}_j^a - \nabla^2 a_j^a + e^2(\delta_{ab} - \delta_{al}\delta_{bl})A_i A_i a_j^b$$

$$+ ef_{lab}\left[(\partial_i A_i)a_j^b + 2A_i \partial_i a_j^b - 2a_j^b - 2a_i^b(\partial_i A_j - \partial_j A_i)\right]$$

$$+ 2e^2(\delta_{ab} - \delta_{al}\delta_{bl})\phi_o^2 a_j^b \quad. \tag{A.4}$$

For $a = 1$ Eq. (A.4) reduces to the free space Maxwell's
equations.  Therefore, modes with $a = 1$ do not see the

background B field and do not produce any instabilities.  On
the other hand, there are six values of a, corresponding to
the other elements of the three U-spin subgroups of SU(4)
that contain a = 1, that do see the background B field.

The modes for different U-spin subgroups are uncoupled,
so we may restrict our attention to the two modes, say $a_j^2$
and $a_j^3$, in a given U-spin subgroup.  It is convenient to
define a complex field

$$a_j \equiv a_j^2 + ia_j^3$$

in terms of which Eq. (A.4) becomes

$$\ddot{a}_j - \nabla^2 a_j + e^2(A^2 + 2\phi_o^2)a_j$$
$$- ie\left[(\nabla\cdot\vec{A})a_j + 2\vec{A}\cdot\nabla a_j - 2a_i B_{ij}\right] = 0 \quad . \tag{A.5}$$

For a background magnetic field in $R^3$ oriented along the three
axis one would have $\vec{A} = \frac{1}{2}\hat{e}_\phi(x_1^2 + x_2^2)^{\frac{1}{2}}B$, where B is the
magnetic field strength.  This would lead to nonvanishing
modes $a_j$ for j = 1,2.  On the other hand, the magnetic field
must diminish in strength as one moves in a direction
orthogonal to $R^3$.  This means the $\partial_i A_j \neq 0$ for $i \neq 1,2,3$.
In turn this means that there will be nonvanishing modes
$a_j$ for j = 4,...,9 in addition to j = 1,2.  The equations
of motion for these modes are

$$\omega^2 a_j = \left[-\nabla^2 - ieB\frac{\partial}{\partial\phi} + e^2(A^2 + 2\phi_o^2)\right]A_j + 2ieB_{ij}a_i \tag{A.6}$$

Among the eight independent solutions to Eq. (A.6) there will
be seven independent growing modes: $a_x + ia_y$ and $a_i + ia_\phi$
for i = 4,...,9.  The eigenvalues for the $a_x + ia_y$ mode are
(10):

$$\omega^2 = k_3^2 + eB(2n - 1)$$

where $k_3$ is the wavenumber along the 3-axis and n is an
integer.  For small $k_3$ and n = 0, $\omega^2 < 0$ indicating

instability.  Similarly for the $a_i + ia_\phi$ mode we find

$$\omega^2 = k^2 + eB(2n + 1) - 3\rho B' + 2e^2\phi_o^2$$

where B' is the derivative of B in a direction orthogonal to $R^3$ and $\rho$ is a scale parameter for the normal vacuum region. In the thin wall approximation $\rho B'$ will exceed B and so we would find instability for small k, n = o at the boundary between the normal and superconducting regions where $\phi_o \to 0$.

Thus we come to the conclusion that if the normal vacuum region is large enough (allowing small k) there will be nonvanishing non-abelian components of the gauge field present. Furthermore, if the normal vacuum region is embedded in a higher dimensional space there will be non-abelian gauge field components pointing outward into the embedding space as well as lying within the normal vacuum region.

NOTICE

This report was prepared as an account of work sponsored by the United States Government. Neither the United States nor the United States Department of Energy, nor any of their employees, nor any of their contractors, subcontractors, or their employees, makes any warranty, express or implied, or assumes any legal liability or responsibility for the accuracy, completeness or usefulness of any  information, apparatus, product or process disclosed, or represents that its use would not infringe privately-owned rights.

Reference to a company or product name does not imply approval or recommendation of the product by the University of California or the U.S. Department of Energy to the exclusion of others that may be suitable.

PHYSICAL STATES AND RENORMALIZED OBSERVABLES
IN QUANTUM FIELD THEORIES WITH EXTERNAL GRAVITY*

S. A. Fulling

Department of Mathematics
Texas A&M University
College Station, Texas


I wish to report on the present status of quantum field theory in curved space-time -- that is, with an external gravitational field.

In such a theory one starts with a prescribed space-time geometry (metric tensor),

$$ds^2 = g_{\mu\nu}dx^\mu dx^\nu,$$

such as that of a homogeneous, isotropic expanding universe,

$$ds^2 = dt^2 - a(t)^2(dx^2 + dy^2 + dz^2).$$

Or the metric might be that of a black hole, or of a collapsing star (as in the famous problem studied by Hawking), or any other model in which the curvature becomes strong enough to affect microscopic processes. On this fixed classical background one has one or more quantized fields satisfying covariant wave equations and commutation relations, such as

$$L\phi \equiv g^{\mu\nu}\nabla_\mu\nabla_\nu\phi + \xi R\phi + m^2\phi = 0,$$

$$[\phi(t,\vec{x}), \pi(t,\vec{y})] = i\delta(\vec{x},\vec{y})$$

---

*Report presented at the Seminar on Quantum Gravity, Moscow, Dec. 5-7, 1978, and at the New Orleans Conference on Quantum Theory and Gravitation, May 23-26, 1979.  Research supported in part by National Science Foundation Grant No. PHY 77-01432.



Copyright © 1980 by Academic Press, Inc.
All rights of reproduction in any form reserved.
ISBN 0-12-473260-7

for a scalar field.  For the time being we study only <u>linear</u> field equations (although, of course, interacting fields are also of interest and
are beginning to be studied).

I  am interested in this subject primarily for its own sake; it raises
many interesting conceptual problems and technical mathematical questions.
But its potential broader significance is twofold:  First, it is believed
to be relevant in astrophysical situations such as the early universe,
black holes, and maybe others.  Second, I believe that we shall have to
understand this before we learn how to quantize gravity; it has somewhat
the same relation to a full quantum gravity as the elementary Schrödinger
equation with a potential has to the full atomic physics of mutually
interacting electrons and radiation.

Thus our immediate aim is to develop a consistent theory of a quantized matter field, $\phi$ ,  in a given background metric,  $g_{\mu\nu}$.  Since direct
experimental verification is rather remote, one hopes that such a theory
will be determined almost uniquely as an extension of general principles
which are already understood -- quantum theory, relativity (special and
general), and the analogy with an external electromagnetic field (which
has received considerable attention in recent years especially in the
Soviet Union).  Of course, the <u>ultimate aim</u> is to understand the effect
of this quantized matter back on the geometry.  The simplest way to try to
do this is the semiclassical approach, where one puts on the right-hand
side of Einstein's equation an expectation value in some state of the
energy-momentum tensor of the quantized fields:

$$R_{\mu\nu} - \frac{1}{2} R\, g_{\mu\nu} = -G \langle T_{\mu\nu} \rangle.$$

The alternative is to quantize the gravitational field itself along with
the others; or perhaps the correct fundamental theory of nature will be
of an entirely new type.  The problem of the dynamics of the gravitational

field is always in the back of our minds as we think about how to do quantum field theory in a given gravitational field.

One might think that a quantum field theory with linear field equations would by now be considered trivial. On the contrary, external-field theories are beset by serious conceptual problems. I think there are basically two of these, which affect each other in a circular manner. The first question is this: After one has generalized the mathematical formalism of quantum field theory to curved space-time, how is that apparatus to be interpreted physically? Things which ought to be observables, such as the electric current and the energy-momentum tensor, are given to us as products of two field operators at the same point in space, and naive attempts to calculate such things lead to meaningless divergent expressions. The standard textbook solution to the divergences of a linear field theory is normal ordering, but that simply does not apply here, partly because of problem No. 2: It is impossible to define a vacuum state, because it is impossible, in general, to interpret the states of the theory in terms of particles. In this theory the field aspect of nature appears to be more fundamental than the particle aspect. (Of course, this assumption is built into our approach, since we start from a field formalism.) Particle phenomena emerge only under certain circumstances, when the space-time is, in some sense or other, almost flat. The particles are very much like the phonons, and so on, of solid state physics, which are precisely defined, stable excitations only in a perfect crystal. When the particle concept becomes imprecise, the states cannot be characterized and classified by their particle content. This leaves us with the problem of what kind of physical content we should look for -- and this leads us back to problem 1, defining observables.

It is hard to overstate the confusion that existed in this subject around 1974, but in the few years following, tremendous progress was made toward putting the theory on a sound foundation. Today, I believe that

the remaining disputes among workers in the field are largely matters of
taste and semantics.  My account of the situation will be very biased
toward my own taste.  First, I will describe how renormalized observables
(especially the energy-momentum tensor) can be defined covariantly and
convincingly; most of the definitive recent work on this problem has been
done by Robert Wald, who talked about it himself in Moscow, so I will
be as brief as I can, before going on to discuss the implications of
this work for the identification of physical states.

The approach to renormalization I am talking about is the most modern
version of the notorious "point-splitting."  It starts from the observation that formally a quantity like the square of the field ought to be
a limit of a product of fields at two _different_ points,

$$\phi(x)^2 = \lim_{y \to x} \phi(x)\phi(y),$$

and that the latter is a well-defined function as long as the points are
spacelike or timelike separated.  The energy-momentum tensor is a sum of
similar terms:

$$T_{\mu\nu}(x) = \lim_{y \to x} \frac{1}{2} g_{\mu}^{\nu'} \, \partial_{\mu}\phi(x) \, \partial_{\nu'}\phi(y) + \dots \; .$$

Also, we note that the expectation value of  $\phi(x)\phi(y)$  in any reasonable
physical state should be a solution of the field equation in both
variables:

$$\langle\phi(x)\phi(y)\rangle \equiv G(x,y), \qquad L_x G = 0 = L_y G.$$

Now it is a classical mathematical fact, due to Hadamard, that a series
solution of  $L_x G = 0$  can be constructed about  $y$,  whose coefficients
depend only on the metric and its derivatives at  $y$:

$$G_{Had}(x,y) = \frac{C_1}{\sigma} + (C_2 m^2 + C_3 \xi R)\, \log \sigma + C_4 R_{\mu\nu} \frac{\nabla_\mu \sigma \nabla_\nu \sigma}{\sigma}$$

$$+ \sigma \log \sigma (C_5 R^2 + C_6 \nabla_\mu \nabla^\nu R + \ldots) + \ldots$$

$$+ \text{terms antisymmetric in } x \text{ and } y \,.$$

Here I have written out a few typical terms. The $C_j$'s are certain

constants, and $\sigma$ is half the square of the (indefinite) geodesic

distance between $x$ and $y$. This series contains only positive powers

of the mass and remains valid for $m = 0$. The well-known DeWitt series

for a Green function of the quantized field carries the expansion on to

terms like

$$\frac{1}{m^2} (C_5 R^2 + C_6 \nabla_\mu \nabla^\mu R + \ldots),$$

but these expressions are irrelevant to a massless field. Nevertheless,

the singular terms of the DeWitt series are the same as those of the

Hadamard series, and they are the proper terms to subtract in renormal-

izing a massless theory. (This is one of the things that was not well

understood until recently.)

Now, if our expectation value $G = \langle \phi\phi \rangle$ has the same short-distance

expansion as $G_{Had}$ except for smooth, nonsingular terms, then a renor-

malized function, $G_{renorm}$, can be obtained by subtracting the singular

terms of $G_{Had}$ from $G$. (The remaining nonsingular terms are not at all

unimportant -- in fact, they contain all the physics of the situation

under study, and our goal is to isolate them from the singular terms,

which are the same for all states and hence carry no physical infor-

mation.) The renormalized

$$\langle \phi(x)^2 \rangle = \lim_{x \to y} G_{renorm}(x,y)$$

is now finite, and in a similar way one arrives at a renormalized

$\langle T_{\mu\nu}(x)\rangle$, which is finite and has a number of properties which testify to its physical correctness:

1. It is conserved: $\nabla_\nu T_\mu^{\ \nu} = 0$. (This property must be _imposed_ in the definition of $T_{\mu\nu}$ (renormalized) to break the ambiguity of the separation into singular and nonsingular parts.)

2. In certain models where everything can be calculated, such as a 2-dimensional model of the Hawking process, the results give a physically plausible picture of the currents of matter in the problem.

3. It is manifestly correct in regions where the space-time is flat; that is, it is the standard energy-momentum tensor of the particles present in the corresponding state of the free field (for example, the particles created in asymptotically static cosmological expansions as calculated by Parker).

4. Its dependence on $g_{\mu\nu}$ is properly causal.

5. It is essentially unique. There is an inherent ambiguity in the result which can be absorbed into a renormalization of coupling constants of fourth-order terms in the gravitational field equation. (As Wald has emphasized, these terms are not to be accepted lightly, however, since they may make the standard solutions of Einstein's equation unstable against perturbations.) Recently Martin Brown has criticised the uniqueness argument on the grounds that there may be local geometrical tensors which are not polynomials in the curvature, and these would introduce further ambiguity. But as far as I know, all the monsters of this kind which have been exhibited are singular in the limit of flat space and hence not viable candidates for inclusion in the renormalized $T_{\mu\nu}$.

Now I turn to the main point of my report. The foregoing discussion suggests that the "physically acceptable" states of a quantum field are precisely the states for which $\langle\phi(x)\phi(y)\rangle$ has the same short-distance singularity structure as $G_{Had}(x,y)$. (This guarantees that the renormalized observables are finite.) I believe that physical states in this sense exist for every space-time model, or at least for every sufficiently smooth, globally hyperbolic metric. (If the metric is not globally hyperbolic, it is not clear what the field commutator ought to be. This is a direction in which the basic theory needs to be extended.) Let me review the progress that has been made toward verifying this conjecture.

1. The conjecture is supported by -- in fact, originated in -- the work of DeWitt and Christensen, Sakharov, Hawking, Dowker, and others, all of which involve, explicitly or implicitly, the replacement of the 4-dimensional indefinite-metric space-time manifold by a compact 4-dimensional manifold with positive definite metric. These approaches are difficult to make mathematically convincing. There must be a more satisfactory way of reaching the same conclusion.

2. Sweeny, Wald, and I have proved that if a state has the desired Hadamard-singularity property at one time (that is, near a Cauchy surface), then it possesses it for all time. This shows immediately that if the universe starts out flat, then the physical Fock-space states of the initial free field evolve into physical states of the field in the later, curved space-time region.

3. This conclusion can probably be extended to any model which is asymptotically _static_ in the past, regardless of the geometry of the static 3-manifold. We expect to prove this by appealing to the same mathematics used in some of the 4-dimensional formalisms I mentioned earlier, namely, the theorems of McKean and Singer, Gilkey, and other mathematicians, relating the density of eigenvalues of the Laplacian to the Green function of the heat equation on a manifold (3-dimensional in our case). Unfortunately, to be generally applicable this work must be reformulated to include noncompact manifolds, and this may be very difficult.

4. The conjecture is true for Robertson-Walker (general Friedmann-type) universes. This has been shown by explicit calculations by Parker, Fulling, Davies, Bunch, and others. An important and surprising point arises here: The effective frequency of oscillation of the normal modes of the field is shifted from the value suggested by the form of the separated field equation, and the physical states turn out to correspond to a decomposition of the solutions into positive and negative frequencies in this new sense. This is not the same decomposition which diagonalizes the instantaneous Hamiltonian of the field theory, except when the geometry is static.

For an analogy, consider the damped oscillator equation,

$$\frac{d^2\psi}{dt^2} + 2\gamma \frac{d\psi}{dt} + \omega^2\gamma = 0,$$

whose solutions have a frequency which is shifted away from $\omega$:

$$\psi = e^{-\gamma t} e^{-i\sqrt{\omega^2 - \gamma^2}\, t} \quad .$$

In the same way, the oscillator with time-dependent frequency,

$$\frac{d^2\psi}{dt^2} + \omega_k(t)^2 \psi_k = 0,$$

has solutions which are approximated by the expression

$$\psi_k \approx \frac{1}{\sqrt{W_k}} e^{-i \int^t W_k dt'}$$

and its complex conjugate, where $W_k^2$ must have an asymptotic expansion whose first term is $w_k^2$, with higher-order terms depending on derivatives of $\omega_k$. These differ from instantaneously positive-frequency functions in the more obvious sense by a nontrivial Bogolubov transformation:

$$\psi_k \underset{t \approx t_0}{\sim} \alpha_k e^{-i\omega_k t} + \beta_k e^{i\omega_k t} \quad .$$

5. Finally, the latter construction can probably be extended to more general models by generalizing the WKB approximation to systems of coupled equations. For the system

$$\frac{d^2\psi_j}{dt^2} + \sum_{k=1}^{N} M_{jk}(t)\psi_k = 0,$$

one has

$$\psi_k \approx \frac{1}{\sqrt{W_k}}\; \vec{e}_k\; e^{-i\int^t W_k dt'}\,,$$

where

$$\vec{e}_k(t) = \text{an eigenvector of } M(t) + \text{higher-order terms}$$

and

$$W_k(t)^2 = \text{an eigenvalue of } M(t) + \text{higher order terms.}$$

There are complications when 2 eigenvalues cross as $t$ changes, which have not yet been resolved. When this project is finished, it will establish the conjecture for the mixmaster universe, where Hu has shown that the field equation separates into finite multiplets of coupled modes.

In the general case, however, infinitely many modes will be coupled together, and one must extend these asymptotic considerations to partial differential equations or to differential equations in Banach spaces.

In summary, the immediately outstanding problems are to reproduce the Hadamard singularity structure from the normal-mode solutions in a general static model (point 3), and to extend the results to time-dependent models by asymptotic analysis of coupled equations (point 5). Research on these problems is in progress.

BIBLIOGRAPHY

I.    Review articles:

C. J. Isham and others, contributions to Eighth Texas Symposium on
Relativistic Astrophysics, Ann. N. Y. Acad. Sci. 302 (1977), Part
III (pp. 114-190).
B. S. DeWitt, Phys. Reports 19, 295 (1975).
L. Parker, in Asymptotic Structure of Space-Time, ed. by F. P.
Esposito and L. Witten, Plenum, New York, 1977, pp. 107-226.


II.   Mathematical formulations:

C. J. Isham, in Differential Geometrical Methods in Mathematical
Physics II (Lecture Notes in Mathematics 676), ed. by K. Bleuler
et al., Springer, Berlin, 1978, pp. 459-512.


III.  Classic quantum calculations of particle creation:

L. Parker, Phys. Rev. 183, 1057 (1969).
S. W. Hawking, Commun. Math. Phys. 43, 199 (1975).


IV.   Energy-momentum tensors and point-splitting:

1.    B. S. DeWitt, Dynamical Theory of Groups and Fields, Gordon and
Breach, New York, 1965.
S. M. Christensen, Phys. Rev. D 14, 2490 (1976); ibid.  17, 946
(1978).

2.    P. C. W. Davies, S. A. Fulling, and W. G. Unruh, Phys. Rev. D 13,
2720 (1976).
P. C. W. Davies and S. A. Fulling, Proc. Roy. Soc. A 354, 59
(1977).
S. A. Fulling, J. Phys. A 10, 917 (1977).
P. C. W. Davies, S. A. Fulling, S. M. Christensen, and T. S. Bunch,
Ann. Phys. (N.Y.) 109, 108 (1977).
T. S. Bunch, S. M. Christensen, and S. A. Fulling, Phys. Rev. D 18,
4435 (1978).

3.    R. M. Wald, Commun. Math. Phys. 54, 1 (1977); Phys. Rev. D 17,
1477 (1978); Ann. Phys. (N.Y.) 110, 472 (1978).
G. T. Horowitz and R. M. Wald, Phys. Rev. D 17, 414 (1978).


V.    "Progress toward verifying the conjecture":

1.    A. D. Sakharov, Teor. Mat. Fiz. 23, 178 (1975) [Theor. Math. Phys.
23, 435].
S. W. Hawking, Commun. Math. Phys. 55, 133 (1977).
J. S. Dowker and R. Critchley, Phys. Rev. D 13, 3224 (1976); ibid.
16, 3390 (1977).
J. S. Dowker and G. Kennedy, J. Phys. A 11, 895 (1978).
See also papers IV.1.

2.    S. A. Fulling, M. Sweeny, and R. M. Wald, Commun. Math. Phys. 63,
257 (1978).

3.   C. Clark, SIAM Review $\underline{9}$, 627 (1967).
R. Balian and C. Bloch, Ann. Phys. (N.Y.) $\underline{60}$, 401 (1970); ibid. $\underline{64}$, 271 (1971); ibid. $\underline{69}$, 76 (1972).
H. P. McKean and I. M. Singer, J. Diff. Geom. $\underline{1}$, 43 (1967).
P. B. Gilkey, J. Diff. Geom. $\underline{10}$, 601 (1975).

4.   L. Parker and S. A. Fulling, Phys. Rev. D $\underline{9}$, 341 (1974).
S. A. Fulling, Gen. Rel. Grav. $\underline{10}$, in press.
T. S. Bunch and P. C. W. Davies, Proc. Roy. Soc. A $\underline{357}$, 381 (1977); ibid. $\underline{360}$, 117 (1978); J. Phys. A $\underline{11}$, 1315 (1978).
T. S. Bunch, Phys. Rev. D $\underline{18}$, 1844 (1978).
N. D. Birrell, Proc. Roy. Soc. A $\underline{361}$, 513 (1978).
See also some of papers IV.2.

5.   B. L. Hu, Phys. Rev. D $\underline{8}$, 1048 (1973).
S. A. Fulling, J. Math. Phys. $\underline{16}$, 875 (1975); ibid. $\underline{20}$, 1202 (1979).
N. M. J. Woodhouse, Phys. Rev. Lett. $\underline{36}$, 999 (1976).

# QUANTUM ASPECTS
# OF GEOMETRODYNAMICS

Maurice J. Duprè[1]

Department of Mathematics
Tulane University
New Orleans, Louisiana

Since the purpose of a conference on quantum gravity is to examine very speculative ideas in view of our present state of knowledge, I would like to consider some aspects of geometrodynamics which seem to be of a rather quantal nature, and which relate to an earlier speculation (2).

I would first like to remark that in the double slit experiment, as pointed out by Dirac, the interference effects are due to photons interfering with themselves (1). Two photons cannot cancel each other out. The quantum interpretation is that the wave function of a single photon is a superposition of wave functions corresponding to the photon passing through each of the slits. Since the photons all come from a single source, this is not unbelievable although it appears strange at first. However, interference can also be observed when two distinct sources send photons to a common photographic plate. In this case, the interpretation is that the wave function of a single photon reaching the plate is a superposition of two wave functions corresponding to the photon coming from each of the sources. It is conceivable that somehow a common control device used to operate each source is somehow responsible for a photon from one source knowing that when the other source is turned on it should then

---

[1]Supported by NSF grant MCS-76-05947A01.



Copyright © 1980 by Academic Press, Inc.
All rights of reproduction in any form reserved.
ISBN 0-12-473260-7

possibly interfere with itself.  The quantum viewpoint seems
to deny the view that the photon actually does come from one
of the sources and not both.  In fact, I have been informed
that interference has been observed in light reaching us from
distinct galaxies, and it is not unreasonable to assume that
photons reaching us from two galaxies A and B producing interference even when A and B are separated by a spacelike interval so they could not interact, and photons leaving these
sources would not interact until close to the earth.  To
demand that the photon is in a superposition of states "from
each source" I think now becomes too unreasonable.  It would
seem more reasonable to view the quantum state of the photon
as merely embodying certain information about the particle,
and admit that the reality is simply not understood in quantum
theory.  The classical electromagnetic wave theory gives a
correct prediction here, so presumably the photons introduced
via quantum electrodynamics would have wave functions of the
right sort, but the naive approach seems to fail.

The point here is that if you look at the foundations of
quantum mechanics say as developed by Jauch, it must be admit-
ted that until symmetry information from the real world, i.e.,
information of a factual geometric nature is introduced, the
quantum theory is merely a general theory of experiment and
information, and could conceivably apply to almost any kind of
system - even one not arising in physics (3).

My own area of primary research interest is C*-algebra
and von Newmann algebra, and the more I study these mathematical systems, the more I become convinced, even though I have
no proof, that almost any mathematical problem or model could
be studied using C*-algebra and von Neumann algebra theory.
In fact I would speculate that possibly it might be proven

that some large class of mathematical problems could be
embedded in the category of C*-algebras in some fashion so
as to turn a large class of mathematical problems into
C*-algebra problems, and thus into problems about operators on
Hilbert space.  In fact the procedure that Jauch uses for
quantum mechanics might simply generalize in some way.  With
this in mind, it is not terribly far fetched to imagine that
there might be some geometrical or classical model of the
universe whose operator theoretic representation gives quantum mechanics in the usual form.

The model which interests me the most on grounds of
mathematical beauty (a perfectly good reason for a mathematician) is the empty universe model of Einstein's general relativity with the vacuum electromagnetic field as the only
source (2).  In this model, I am imagining particles as being
represented by small topological tangles in the structure of
spacetime.  At this level of generality, an enormous zoo of
particles is allowed, and it seems that there should be ample
room to include all the known elementary particles.  I am not
saying I can prove this mathematically, but it doesn't seem to
me unreasonable.  For all I know, however, the model of an
electron in such a scheme might be quite complicated.

Suppose we try to be more specific and make some simpli-
fying assumptions.  We can imagine a space time of the form
$S \times \mathbb{R}$ where $S$ is a Riemannian 3-manifold obtained by taking a
copy of $S^3$, taking $10^{80}$ points on $S^3$, cutting out a small 3-
disc around each point and connecting them in pairs by handles
of the form $D^2 \times I$.  Thus our space $S$ has a 2nd cohomology
group with about $10^{80}$ generators and thus so does $M = S \times \mathbb{R}$ .
Now, as discovered by Rainich and improved upon somewhat by
Misner and Wheeler, provided certain conditions are satisfied

by the Ricci curvature, there is a unique electromagnetic
field F on M, unique up to multiple by complex number ($\Lambda^2 M$
has complex structure given by the Hodge *-operator) of
modulus one (4).  Now in this model I am thinking of the
mouth of one of these handles "or wormholes" as representing a
particle.  To the outsider, the particle appears to have mass
because the space is curved around the outside of the mouth
even though the scalar curvature vanishes.  If $S_0$ is an embedded 2-sphere surrounding one of these wormhole mouths, then

$$\int_{S_0} (*F) = q_0$$

gives the apparent charge to an outside observer.  If we want
to think of these as representing elementary particles, then
we should have $\int_{S_0} (*F) = \pm\, e$, when e is the electron charge.  Put
another way, we have $\int_{S_0} *\frac{F}{e} = \pm\, 1$, or that for any smooth singular integral 2-cycle C on M

$$\int_C *\frac{F}{e} \; \epsilon \; Z, \text{ the ring of integers.}$$

This is simply a Bohr-Sommerfeld quantization condition, which
certainly seems justified by observation.  The condition is
simply that $\frac{F}{e}$ corresponds to an integral cohomology class
under the de Rham isomorphism.  Once that is postulated, we
can ask ourselves how these "particles" interact.  In the
limit as wormhole radii $\to 0$, the particles become singularities and thus travel along geodesics.  However, suppose we
consider two positively charged particles $P_1$ and $P_2$.  Suppose
the smallest sphere $S_1$ surrounding the wormhole mouth, $P_1$, has
area $A_1$.  By spherical symmetry we can argue (assuming the
particles to be roughly uniformly distributed around $P_1$, we
expect at least approximate spherical symmetry) that the field

F on $S_1$ has roughly constant outward normal component, i.e.,

$$1 = \int_{S_1} *(F) = A_1 E_1, \text{ taking } e = 1,$$

where $E_1$ is the electric field normal to $S_1$. Thus $E_1 = \frac{1}{A_1}$, and we see that the electric field strength has $\frac{1}{A_1}$, as an upper bound. In particular, as the curvature in the wormhole "throat" should be very large, it would seem that two positively charged particles when close enough would strongly attract each other. That is, the gravitational attraction should overpower the electric repulsion at extremely short distances. Moreover, if $P_3$ is a particle with charge zero, then when $P_1$ and $P_3$ are close enough, the gravitational attraction would hold them tightly together and thus increase the curvature even more, which would make it even easier for $P_2$ to be held in tight if it were able to get sufficiently close. In other words, we would seem to have a possible model for an atomic nucleus, which seems to indicate a reason why large nuclei require neutrons for stability, whereas very small nuclei do not. Can there possibly be a connection with this model and actual quantum mechanics? We could try to carry out the Jauch type program for M instead of physical reality. In particular, in a neighborhood of an event $x_0 \in$ M, or wormhole mouth, we imagine that the laboratory physicist sets up a coordinate system $\mathbb{R}^4$ in his mind - he does this macroscopically, ignoring all the complicated structure in the same way a cosmological model of the universe would ignore the ordinary scale wrinkles in the spacetime structure. But the result is, he is coordinitizing a region of spacetime which cannot be coordinitized because of the topological complications, and he is attempting to force the measurement process

to produce exact locations for particles which are non-local
on the microscopic scale - obviously the position of a wormhole mouth can have no precise meaning.  It is now not hard
to imagine that in an attempt to measure the position of one
of these particles - which can only be done by letting it
interact with others of the particles - many fluctuations
would occur due to his inability to adequately control the
interactions and initial conditions.  If a is an observable,
the natural thing to do then is let p(a) be the expectation
value of measuring a when the system has state p.  We can talk
about yes-no questions concerning the model as "projected"
onto the "coordinate system" $\mathbb{R}^4$ used in the laboratory, and
use these yes-no questions to build arbitrary observables.
When we represent the lattice of yes-no questions by projection operators on Hilbert space, the observables become integrals of spectral measures, i.e., self-adjoint operators, and
we have roughly a standard quantum formulation.  For instance,
for a "free" particle we expect the generator of the dynamical
group to roughly be rotationally symmetric when viewed classically.  Or, relativistically Lorentz invariance should be
applicable to arrive at a relativistic quantum theory.  Moreover, the uncertainty in measurement should be related to the
"radius," r, of the wormhole mouths which should appear as a
relation between r, $\hbar$, c, m where m is the apparent mass of
the particles.

In order to study the geometrodynamic models having
topological complications it seems that a cobordism theory
approach will be necessary.  Possibly "solutions" for two
particle systems could be worked out with an asymptotically
nice background, and then several different solutions might
be glued together.  In any case, it would seem that to study

such models, radically different methods are necessary, and
that methods which are of proven success in differential
topology should be tried first (5).


ACKNOWLEDGMENTS


I would like to thank the Department of Mathematical
Sciences at Montana State University for providing facilities
during the summer of 1979.


REFERENCES


1.   Dirac, P. A. M., "The Principles of Quantum Mechanics,"
     fourth edition, Oxford University Press, Oxford, (1978).
2.   Dupre, M. J. in "Mathematical Foundations of Quantum
     Theory" (R. Marlow, ed.), p. 339.  Academic Press, New
     York, 1978.
3.   Jauch, J., "Foundations of Quantum Mechanics," Addison-
     Wesley, Reading, Mass., 1968.
4.   Wheeler, J. A., "Geometrodynamics," Academic Press, New
     York, 1962.
5.   Yodzis, P., Commun. Math. Phys., 26 (1972), p.39.

# ON THE INTERPRETATION
# OF QUANTUM MECHANICAL SCATTERING MEASUREMENTS

Jerome A. Goldstein*

Department of Mathematics
Tulane University
New Orleans, Louisiana

## I.  INTRODUCTION

Let  A  be a bounded linear operator on a complex Hilbert space  $H$.
In order to know the operator  A , it is enough to know all the matrix
elements  $\langle A\phi,\psi\rangle$  as  $\phi$  and  $\psi$  range over  $H$.  Sometimes it suffices to
know less.  For example, if  A  is a nonnegative self-adjoint operator,
suppose we know all the values of the quadratic form

$$Q_A(\phi) = \langle A\phi,\phi\rangle = \|A^{1/2}\phi\|^2 \ ,$$

$\phi \in H$.  Then by the polarization identity
$4\langle\phi,\psi\rangle = \|\phi + \psi\|^2 - \|\phi - \psi\|^2 + i\|\phi + i\psi\|^2 - i\|\phi - i\psi\|^2$ , we can determine
$\langle A\phi,\psi\rangle$  for all  $\phi$ , $\psi$ ,  and so we know  A.

In the Hilbert space description of quantum mechanics one generally
measures the amplitudes  $|\langle A\phi,\psi\rangle|^2$  rather than the matrix elements
$\langle A\phi,\psi\rangle$.  The purpose of this paper is to show how measurements of such
amplitudes determine a scattering operator  S  under the assumption that
S  is unitary.  We shall deal separately with the questions of measurements that one can make in principle and measurements that one makes in

---

*Partially supported by an N.S.F. grant.



Copyright © 1980 by Academic Press, Inc.
All rights of reproduction in any form reserved.
ISBN 0-12-473260-7

practice.

It was a great pleasure to present this paper to an audience including John Archibald Wheeler. Not only did Professor Wheeler invent the S-matrix in 1937 [10] (see also Heisenberg [3]), but he also has some very interesting recent thoughts concerning scattering theory [11].

Over the years I have learned a great deal from various colleagues through conversations and correspondence about the interpretation of measurements in scattering theory. For their patience and help I thank Paul Chernoff, Mike Frame, Karl Hofmann, Fran Narcowich, and Barry Simon.

## II. A PRELIMINARY LEMMA

We begin with a bit of folklore; namely, knowing $|<A\phi,\psi>|^2$ for all $\phi,\psi \in H$ means we know $A$ up to a phase shift, i.e. up to a complex multiple of modulus 1. A more precise statement is as follows.

LEMMA 1*. *Let* $A_1$ , $A_2$ *be bounded linear operators on a (complex) Hilbert space* $H$. *If*

$$|<A_1\phi,\psi>|^2 = |<A_2\phi,\psi>|^2$$

*for all* $\phi,\psi \in H$ , *then* $A_1 = cA_2$ *for some complex number* $c$ *with* $|c| = 1$.

PROOF. Fix $\phi \in H$. Knowledge of $|<A_1\phi,\psi>|^2$ for all $\psi \in H$ determines both $\|A_1\phi\|$ and $\{A_1\phi\}^\perp$. It follows that there is a complex number $c_\phi$ with $|c_\phi| = 1$ such that $A_1\phi = c_\phi A_2\phi$. This is true for all $\phi \in H$.

Now let $\{e_\alpha \mid \alpha \in J_1\}$ be an orthonormal basis for $N(A_1)$ = the null space of $A_1$ , and let $\{e_\alpha \mid \alpha \in J_2\}$ be an orthonormal basis for $N(A_1)^\perp$.

---

*This lemma must certainly be well-known, but we have been unable to find a reference.

Thus $J_1$ and $J_2$ are disjoint, and $\{e_\alpha \mid \alpha \in J_1 \cup J_2\}$ is an orthonormal basis for $H$.

If $J_2$ is empty, then $A_1 = A_2 = 0$ and we are done. If not, pick and fix $\beta \in J_2$. Choose $c$ with $|c| = 1$ such that $A_1 e_\beta = cA_2 e_\beta$. It suffices to show that $A_1 e_\alpha = cA_2 e_\alpha$ for all $\alpha \in J_2 \backslash \{\beta\}$ (hence for all $\alpha \in J_1 \cup J_2$ since this automatically holds for the other $\alpha$'s). By the first part of the proof, there are numbers $d_1$, $d_2$ with $|d_1| = |d_2| = 1$ so that $A_1 e_\alpha = d_1 A_2 e_\alpha$, $A_1(e_\alpha + e_\beta) = d_2 A_2 (e_\alpha + e_\beta)$. But then $A_1(e_\alpha + e_\beta) = d_1 A_2 e_\alpha + cA_2 e_\beta = d_2 A_2 e_\alpha + d_2 A_2 e_\beta$. Since $A_2 e_\alpha$ and $A_2 e_\beta$ are linearly independent (otherwise some linear combination of $e_\alpha$ and $e_\beta$ would belong to $N(A_2) = N(A_1)$) we must have $d_1 = d_2 = c$. $\square$

III.   MEASUREMENTS IN PRINCIPLE

Let $S$ be a scattering operator, which we assume to be unitary. (Cf. e.g. [4], [8].) Then for $\phi, \psi \in H$, in a scattering experiment we can *in principle* simultaneously measure the amplitudes $|<R\phi,\psi>|^2$ for a finite commuting family of operators $R$ which commute with $S$. We first do this for one operator: $T = S - I$.

THEOREM 1.  *Let* $I + T_1$, $I + T_2$ *both be unitary, and suppose*

$$|<T_1\phi,\psi>|^2 = |<T_2\phi,\psi>|^2$$

*holds for all* $\phi, \psi \in H$. *Then either*

    (i)  $T_1 = T_2$

*or else*

    (ii)  $T_1 = \omega P$, $T_2 = \bar{\omega} P$ *where* $P$ *is a self-adjoint projection and* $|\omega + 1|^2 = 1$.

PROOF.  Write $T$ for $T_1$. By Lemma 1, $T_2 = cT$ where $|c| = 1$. By hypothesis, $S_\alpha = I + \alpha T$ is unitary for $\alpha = 1, c$. $S_\alpha S_\alpha^* = I$ implies

$$I + \alpha T + \overline{\alpha} T^* + |\alpha|^2 T T^* = I \ ,$$

whence $2\mathrm{Re}(\alpha T) = -T T^*$ (where $2\mathrm{Re}(L) = L + L^*$). Consequently, taking $\alpha = 1$ and $\alpha = c$ we get

$$\mathrm{Re}((1 - c)T) = 0.$$

Assume $c \neq 1$, i.e. (i) does not hold. Then $(1 - c)T = iR$ where $R^* = R$. Thus $I + \beta R$ is unitary for some complex number $\beta$ and some nonzero self-adjoint operator $R$. $(I + \beta R)(L + \overline{\beta} R) = I$ gives $2(\mathrm{Re}\ \beta)R = -|\beta|^2 R^2$. $R = 0$ on $N(R)$ and $R$ is one-to-one on $N(R)^{\perp}$; so on $N(R)^{\perp}$ we get $2(\mathrm{Re}\ \beta)I = -|\beta|^2 R$. It follows that $R = \gamma P$ where $P$ is the orthogonal projection onto $N(R)^{\perp}$. Next, for $P$ a self-adjoint projection, $I + \delta P$ is unitary if and only if $2\mathrm{Re}(\delta) = -|\delta|^2$ if and only if $|1 + \delta|^2 = 1$ from this and from $\|T_2\phi\| = \|T_1\phi\|$ for all $\phi$, it follows that $T_1$, $T_2$ are as advertised in (ii). $\square$

According to the above theorem, from measuring $|<T\phi,\psi>|^2$ and knowing the unitarity of $S = I + T$, either we know $T$ or else $T$ has a trivial form, and this latter case does not seem to occur in the applications. But even when $T$ has the trivial form $T = \omega P$, $S$ can be determined.

COROLLARY 1. *We know both $P$ and $\omega$.*

PROOF. Measuring $|<T\phi,\psi>|^2 = |\omega|^2 |<P\phi,\psi>|^2$ for all $\phi$, $\psi$ tells us what $P$ and $|\omega|$ are. Thus $\omega$ lies on two circles, one centered at the origin of known radius $|\omega|$ and the other centered at $-1$ and of unit radius. Let $\omega_1$ be a point of intersection of these circles. Then $\omega$ is $\omega_1$ or $\overline{\omega}_1$. By simultaneously measuring $T - \omega_1 P$ (which is $0$ or $-2i(\mathrm{Im}\ \omega)P$) we determine $\omega$. $\square$

To conclude this discussion we write down the general unitary operator $S$ on $\mathbb{C}^2$ of the form $S = I + \omega P$ where $P = P^* = P^2$ and $|\omega + 1|^2 = 1$. Either $P = I$ or else for some $\alpha$ in the extended complex plane,

$$S = S_{\alpha,\omega} \equiv \begin{pmatrix} 1 + \dfrac{\omega}{1 + |\alpha|^2} & \dfrac{\omega\alpha}{1 + |\alpha|^2} \\[2em] \dfrac{\omega\bar{\alpha}}{1 + |\alpha|^2} & 1 + \dfrac{\omega|\alpha|^2}{1 + |\alpha|^2} \end{pmatrix}.$$

## IV.  MEASUREMENTS IN PRACTICE

Suppose that the scattering operator $S$ is determined by the free Hamiltonian $H_0 = -\Delta = -\sum_{j=1}^{n} \partial^2/\partial x_j^2$ on $H = L^2(\mathbb{R}^n)$ and the perturbed Hamiltonian $H = H_0 + V$, where $V$ is the multiplication operator $(V\phi)(x) = V(x)\phi(x)$, and the real-valued function $V$ is such that $S$ exists and is unitary on $H$. Sufficient for this is that $V$ is a short range potential in the sense that $V = V_1 + V_2$, $V_1 \in L^\infty(\mathbb{R}^n)$, $V_2 \in L^p(\mathbb{R}^n)$ $(p \geq 2, p > n/2)$, and $V(x) = O(|x|^{-1-\varepsilon})$ as $|x| \to \infty$ for some $\varepsilon > 0$. (See Agmon [1] or Kato and Kuroda [6].)

We now turn to the direct integral decomposition $S = \theta \int_0^\infty S(\lambda)\, d\lambda$ which comes from viewing $L^2(\mathbb{R}^n)$ as $L^2([0,\infty)) \otimes L^2(S^{n-1})$ where $S^{n-1} = \{x \in \mathbb{R}^n \mid |x| = 1\}$. This decomposition is a consequence of conservation of energy; the S-matrix preserves the particle's energy. To be more precise, with the aid of the Fourier transform which diagonalizes $H_0$, we get a unitary correspondence from $L^2(\mathbb{R}^n)$ to $L^2([0,\infty);L^2(S^{n-1}))$ so that if $\phi \in L^2(\mathbb{R}^n)$ corresponds to $\{\phi(\lambda) \mid \lambda \geq 0\}$, $\phi(\lambda) \in L^2(S^{n-1})$, in symbols,

$$\phi \longleftrightarrow \{\phi(\lambda)\},$$

then

$$H_0 \ \phi \ \longleftrightarrow \ \{\lambda \phi(\lambda)\}$$

and

$$S\phi \ \longleftrightarrow \ \{S(\lambda) \ \phi(\lambda)\}$$

where for each fixed $\lambda > 0$ , $S(\lambda)$ is a unitary operator on $L^2(S^{n-1})$. See [5] for more details on this. The S-*matrix* $S(\lambda)$ is typically of the form $I + T(\lambda)$ where $T(\lambda)$ is an integral operator.

Roughly speaking, what one actually measures in scattering experiments are amplitudes of matrix elements such as $|<S(\lambda)\psi_1,\psi_2>|^2$ for fixed $\lambda > 0$ and certain (but not all) states $\psi_1$ , $\psi_2$ in $L^2(S^{n-1})$ , namely, plane waves coming in and going out. Before making this more precise we state somewhat informally the main result of this section.

THEOREM 2. *Let* $\lambda > 0$ *and let* $t_\lambda$ , $\alpha_\lambda$ , $\beta_\lambda$ *be respectively the kernels of the integral operators* $T(\lambda) = S(\lambda) - I$ , $\mathrm{Re}\ T(\lambda)$ , *and* $\mathrm{Im}\ T(\lambda)$ ; *thus* $t_\lambda = \alpha_\lambda + i\beta_\lambda$. *Then scattering experiments can be performed to determine* $\alpha_\lambda$ *and* $|\beta_\lambda|$ , *but not* $\beta_\lambda$ *in general. Let* $C_1$ *be the set of* $\{z \in \mathbb{C} \mid |z| = 1\}$ *valued mappings* c *on* $S^{n-1} \times S^{n-1}$ *such that changing* $\beta_\lambda$ *to* $c\beta_\lambda$ *gives rise to the same data in scattering experiments. Then when* $\alpha_\lambda$ *and* $\beta_\lambda$ *are real, the nonempty set* $C_1$ *has an even number of members in that* $c \in C_1$ *implies* $-c \in C_1$. $C_1$ *will consist of only two members (viz.* $c \equiv 1$ *and* $c \equiv -1$) *in certain situations, for instance if* $\alpha_\lambda$ *and* $\beta_\lambda$ *are real and if* $t_\lambda$ *is a* $C^\infty$ *function on* $S^{n-1} \times S^{n-1}$ , *all of whose zeros are of finite order.*

It is not clear that the experimenter can really make the measurements to determine all that is claimed in the above theorem. Indeed, human experimenters can only make finitely many measurements under any circumstances. But even if not all these measurements are possible in practice, the set-up of Theorem 2 seems to be closer to experimental reality than is the set-up of Theorem 1.

We now begin the proof of Theorem 2. Changing the notation slightly, for added generality and clarity, let $(\Omega, \Sigma, \mu)$ be an arbitrary measure space, and on $H = L^2(\Omega, \Sigma, \mu)$ let $T_1$ be an integral operator defined by

$$(T_1 \phi)(x) = \int_\Omega t(x,y) \; \phi(y) \; \mu(dy).$$

Writing, as usual, $\mathrm{Re}\, T_1 = \frac{1}{2}(T_1 + T_1{}^*)$ , $\mathrm{Im}\, T_1 = \frac{1}{2i}(T_1 - T_1{}^*)$ , we have the following: $T_1 = \mathrm{Re}\, T_1 + i\,\mathrm{Im}\, T_1$ , and $\mathrm{Re}\, T_1$ [resp. $\mathrm{Im}\, T_1$] is an integral operator with kernel

$$\alpha(x,y) \equiv \frac{1}{2}(t(x,y) + \overline{t(y,x)})$$

$$[\text{resp.} \quad \beta(x,y) \equiv \frac{1}{2i}(t(x,y) - \overline{t(y,x)})] \; .$$

Also, $t = \alpha + i\beta$ ; and if $I + T_1$ is unitary, then $\mathrm{Re}\, T_1$ and $\mathrm{Im}\, T_1$ are commuting self-adjoint operators.

We suppose that we simultaneously measure $|\langle(\mathrm{Re}\, T_1)\phi, \psi\rangle|^2$ and $|\langle(\mathrm{Im}\, T_1)\phi, \psi\rangle|^2$ not for all $\phi$ , $\psi$ but rather for $\phi$ , $\psi$ corresponding to incoming and outgoing plane waves. We interpret this to mean that we can measure

$$|\langle(\mathrm{Re}\, T_1)\delta_x, \delta_y\rangle|^2 = |\alpha(x,y)|^2 \; ,$$

$$|\langle(\mathrm{Im}\, T_1)\delta_x, \delta_y\rangle|^2 = |\beta(x,y)|^2$$

for (almost) all $x,y \in \Omega$. Here and in what follows, the delta functions are not bonafide vectors in $H$ (unless $\mu$ has atoms), but it is clear how to approximate them and how to convert formal arguments involving them into rigorous ones.

Thus we measure $|\alpha|$ , $|\beta|$ , where $t = \alpha + i\beta$ is sought. The ambiguous phase shift problem under discussion thus becomes: *Knowing*

$|\alpha|$ , $|\beta|$ , *and the unitarity of* $I + T_1$ *where the kernel of* $T_1$ *is*

$t = \alpha + i\beta$ , *do we know* $t$ ?

The answer is that *we know* $\alpha$ *but we do not know* $\beta$. Here is why.

LEMMA 2. *Let* $I + T_1$ *be unitary. Let* $C$ *be any self-adjoint operator such that* $C^2 = (Im\ T_1)^2$ *and* $[Re\ T_1, C] = 0$. *Then* $I + Re\ T_1 + iC$ *is unitary. Conversely, if* $I + T_2$ *is unitary and* $(Re\ T_1)^2 = (Re\ T_2)^2$ , $(Im\ T_1)^2 = (Im\ T_2)^2$ , *then* $Re\ T_1 = Re\ T_2$ *and* $[Re\ T_1, Im\ T_2] = 0$.

PROOF.  $(I + Re\ T_1 + iC)^* (I + Re\ T_1 + iC)$

$$= (I + Re\ T_1)^2 + C^2 \qquad\qquad \text{since} \quad [Re\ T_1, C] = 0$$

$$= (I + Re\ T_1)^2 + (Im\ T_1)^2 = I$$

since $I + T_1$ is unitary and $[Re\ T_1,\ Im\ T_1] = 0$.

For the converse,

$$I = (I + T_j)^* (I + T_j) = I + 2\ Re\ T_j + (Re\ T_j)^2 + (Im\ T_j)^2 \quad \text{for} \quad j = 1, 2.$$

Since $(Re\ T_1)^2 = (Re\ T_2)^2$ and $(Im\ T_1)^2 = (Im\ T_2)^2$ we conclude that $Re\ T_1 = Re\ T_2$. Finally, $0 = [Re\ T_2,\ Im\ T_2] = [Re\ T_1,\ Im\ T_2]$. $\square$

Thus $\alpha$ is known and is determined by the nonpositive self-adjoint operator $Re\ T_1$.

To describe the class of possible solutions of the ambiguous phase shift problem, we introduce some notation. Let $\mathcal{C}$ be the class of measurable functions from $\Omega \times \Omega$ to $\{z \in \mathbb{C} \mid |z| = 1\}$. Let $\beta$ be the kernel we are trying to measure. By Lemma 2 and the discussion preceding it, the set of solutions $\beta_1$ of the ambiguous phase shift problem is $\{\beta_1 = c\beta \mid c \in \mathcal{C}_1\}$ where

$$(1) \quad \begin{cases} \mathcal{C}_1 = \{c \in \mathcal{C} \mid \text{For } C\phi(x) = \displaystyle\int_\Omega c(x,y)\ \beta(x,y)\ \phi(y)\ \mu(dy)\ , \\[2ex] \qquad\qquad [C, Re\ T_1] = 0 \text{ and } C^2 = (Im\ T_1)^2 \}\ . \end{cases}$$

Using the above notation,

$$C^2 \phi(x) = \int_\Omega \int_\Omega c(x,z)\, c(z,y)\, \beta(x,z)\, \beta(z,y)\, \phi(y)\, \mu(dy)\, \mu(dz) \ , \quad \text{so the}$$

condition that $C^2 = (\text{Im } T_1)^2$ becomes

$$(2) \quad \begin{cases} \displaystyle\int_\Omega c(x,z)\, c(z,y)\, \beta(x,z)\, \beta(z,y)\, \mu(dz) \\[2mm] = \displaystyle\int_\Omega \beta(x,z)\, \beta(z,y)\, \mu(dz) \quad \text{for (almost) all} \quad x,y \in \Omega \ . \end{cases}$$

The condition $[C, \text{Re } T_1] = 0$ becomes

$$(3) \quad \begin{cases} \displaystyle\int_\Omega \alpha(x,z)\, c(z,y)\, \beta(z,y)\, \mu(dz) \\[2mm] = \displaystyle\int_\Omega \alpha(x,y)\, c(x,z)\, \beta(x,z)\, \mu(dz) \quad \text{for (almost) all} \quad x,y \in \Omega. \end{cases}$$

Thus we have proved

LEMMA 3. *The set of solutions* $\beta_1$ *of the ambiguous phase shift problem is* $\{\beta_1 = c\beta \mid \beta \in C_1\}$. *Moreover, for* $c \in C$ , $c \in C_1$ *if and only if* (2) *and* (3) *hold.*

Note that condition (3) is needed for the (unitarity, hence) normality of $I + \text{Re } T_1 + iC$. We remark that it is an easy exercise to show that an integral operator $K$ with kernel $k$ is normal if and only if for (almost) all $x,y,z \in \Omega$ ,

$$k(x,y)\, \overline{k(z,y)} = \overline{k(y,x)}\, k(y,z)$$

holds. Thus multiplying $k$ by $c \in C$ can destroy normality.

Conditions (2) and (3) are complicated to verify. For $\mu$ purely atomic and $\alpha(x,y) = \beta(x,y) = 0$ for $x \neq y$ , then *any* real-valued $c \in C$ belongs to $C_1$ (see (1)). On the other hand, if $\Omega$ is a connected topological space, $\Omega$ is the support of $\mu$ , $\alpha$ and $\beta$ are real, and $\beta$ is continuous and never zero on $\Omega$ , then $C_1 = \{c \equiv +1 , c \equiv -1\}$ has only two members.

Decompose $\alpha$ and $\beta$ into their real and imaginary parts as

$\alpha = \alpha_1 + i\,\alpha_2$ , $\beta = \beta_1 + i\,\beta_2$. Then $t = \alpha + i\beta = \alpha_1 - \beta_2 + i(\alpha_2 + \beta_1)$.
We know $|\alpha|^2 = \alpha_1^2 + \alpha_2^2$ , $|\beta|^2 = \beta_1^2 + \beta_2^2$ , and

$$|t|^2 = |\alpha|^2 + |\beta|^2 + 2(\alpha_2\beta_1 - \alpha_1\beta_2).$$

Thus we can measure $\alpha_2\beta_1 - \alpha_1\beta_2$. Note also that if $k$ is the kernel of
a self-adjoint integral operator, then $k(x,x) = \langle T_1\delta_x, \delta_x\rangle$ is real for
all $x \in \Omega$. Very often self-adjoint integral operators have real kernels.
Thus suppose $\alpha_2 = 0$. Then $\alpha = \alpha_1$ is real and $\alpha\beta_2$ is known. Clearly
$c$ can be real on the zero set of $\beta_2$ without changing $\alpha\beta_2$. On the set
$E_1$ where $c$ is nonreal, write $c = c_1 + ic_2$ ($c_2 \neq 0$ on $E_1$). Then
$\mathrm{Im}(c\beta) = c_1\beta_2 + c_2\beta_1$ , which equals $\beta_2$ on $E_3 = E_1 \cap E_2$ , where $E_2$ is
the complement of the zero set of $\alpha$. This imposes many restrictions on
the possible values of $c$. Indeed, the system of equations
$c_1\beta_2 + c_2\beta_1 = \beta_2$ , $c_1^2 + c_2^2 = 1$ will often have a unique solution for
$c = c_1 + ic_2$ , namely $c = 1$. However, assuming that $\alpha$ and $\beta$ are both
real, only real-valued solutions of $c$ are possible. In particular, in
this case, if $c \in C_1$ then $-c \in C_1$ , so the nonuniqueness inherent in a
global sign change in the kernel of the imaginary part of the integral
operator is always present in the ambiguous phase shift problem.

In a scattering experiment one measures quantities such as
$|\langle S(\lambda)\psi_1, \psi_2\rangle|^2$ where $\psi_1$ is a plane wave coming in at one angle and $\psi_2$
is a plane wave going out at another angle. These two waves determine a
plane, and we make all possible measurements in this plane before proceed-
ing to other planes. Thus the relevant measure space becomes the unit
circle $\mathbb{T} \equiv \{z \in \mathbb{C} \mid |z| = 1\}$ , which we regard as a compact abelian group.

More generally, suppose that $\Omega$ is a compact abelian group (under $\cdot$),
that $\mu$ is Haar measure on $\Omega$ , and that the kernel $\beta(x,y)$ depends only

on $xy^{-1}$ , so that $\mathrm{Im}\ T_1$ is a convolution integral operator. The

condition that $C^2 = (\mathrm{Im}\ T_1)^2$ is expressed most easily in terms of the

Fourier transform. The kernel of $C^2$ is $c\beta * c\beta$ , which necessarily

equals $\beta * \beta$. In the Fourier transform representation this becomes

$$(4) \qquad \hat{c\beta}^2 = \hat{\beta}^2 .$$

Returning to $\mathbb{T}$ , define the Sobolev spaces as usual to be

$$W^{k,2}(\mathbb{T}) = \{f \mid f^{(m)} \in L^2(\mathbb{T}) , m = 0,\ldots,k\} ;$$

thus $f \in W^{k,2}(\mathbb{T})$ if and only if $f$ and its first $k$ distributional

derivatives are all square integrable on $\mathbb{T}$. (When differentiating

functions on $\mathbb{T}$ , it is convenient to regard $\mathbb{T}$ as the interval $[0,2\pi)$

under addition modulo $2\pi$.) The smoothness of $f$ determines the rate of

decay of its Fourier coefficients and vice-versa. More precisely

$f \in W^{2,k}(\mathbb{T})$ if and only if

$$\sum_{n=-\infty}^{\infty} |n^k\ \hat{f}(n)|^2 < \infty.$$

This implies $\hat{f}(n) = o(|n|^{-k})$ and is implied by $\hat{f}(n) = O(|n|^{-k-a})$ for

$a > 1/2$. In any event, (4) implies that $c\beta \in W^{k,2}(\mathbb{T})$ if and only if

$\beta \in W^{k,2}(\mathbb{T})$. Thus changing $\beta$ by multiplying it by an allowable

$c(\in C_1)$ cannot destroy differentiability. Thus, in particular, if $\alpha$ and

$\beta$ are real, if $\beta \in W^{k,2}(\mathbb{T})$ , and if $\beta$ has no zero of order* greater

than or equal to $k$ , then necessarily $C_1 = \{c \equiv +1, c \equiv -1\}$. Still

assuming that $\alpha$ and $\beta$ are real, for $\beta \in C^{\infty}(\mathbb{T})$ we have

$C_1 = \{c \equiv +1, c \equiv -1\}$ provided that all the zeros of $\beta$ are of finite

order. These last two sentences justify the last assertion of Theorem 2,

---

<sup></sup>
* $\beta$ has a zero of order $m$ at $x_0$ means $\beta^{(j)}(x_0) = 0$ for $j = 0,$
$\ldots,m-1$ and $\beta^{(m)}(x_0) \neq 0$.

the proof of which is now complete.  □

For a discussion of a different but related ambiguous phase problem, see [7, pp.160,161]. For references see [7, p.172].

Let now $\Omega$ be a locally compact abelian group and let $\hat{\Omega}$ be its dual group. The Fourier transform $f \longrightarrow \hat{f}$ is a unitary mapping from $L^2(\Omega)$ to $L^2(\hat{\Omega})$ (the normalized Haar measures being suppressed from the notation).

PROBLEM.  *Let*

$$C_{\mathbb{R}} = \{c \mid c : \Omega \rightarrow \{-1,+1\}\} \ , \quad C_{\mathbb{C}} = \{c \mid c : \Omega \rightarrow \{z \in \mathbb{C} \mid |z| = 1\}\} \ .$$

*Given* $\beta \in L^2(\Omega)$ *, for which* c *in* $C_{\mathbb{R}}$ *or* $C_{\mathbb{C}}$ *is it true that*

$$|\hat{\beta}| = |\widehat{c\beta}|$$

*on* $\hat{\Omega}$ *?*

$C_{\mathbb{R}}$ [resp. $C_{\mathbb{C}}$] is the class to consider when we are concerned with real-[resp. complex-] valued functions on $\Omega$. For c in $C_{\mathbb{R}}$ or $C_{\mathbb{C}}$ we have necessarily $|\beta| = |c\beta|$ on $\Omega$. Clearly constant functions in $C_{\mathbb{C}}$ or $C_{\mathbb{R}}$ are always solutions.

Let $\Omega = \mathbb{R}^3$ (or more generally $\mathbb{R}^n$). Choose unit vectors $\beta$, $\gamma$ in $L^2(\mathbb{R}^3)$ such that $|\beta|^2 = |\gamma|^2$ , $|\hat{\beta}|^2 = |\hat{\gamma}|^2$. Then $\beta$, $\gamma$ can be thought of as describing the wave functions of spinless nonrelativistic quantum mechanical particles, and these wave functions have the same position probability density $(|\beta|^2)$ and momentum probability density $(|\hat{\beta}|^2)$. Does then $\gamma = c\beta$ for some constant c of absolute value one ? This is a much studied question, the answer to which is no in general and yes in certain cases. (Recent papers on this topic are [2], [9]; [2] has further references.) Intuitively, the information contained in the wave function $\beta$ includes the joint distribution for the position and velocity.

In specifying the marginal distributions of the position and momentum

random variables, one thing missing is the correlation between the two.

Thus the Problem stated above is a significant problem for $\Omega = \mathbb{R}^n$

for the purpose of interpreting probabilistically the quantum mechanical

wave function.  For the circle group $\Omega = \mathbb{T}$ , the Problem is significant

for the purpose of interpreting the connections between experiments and

the S-matrix.  Conceivably choosing $\Omega$ to be another group could lead to

results having a quantum mechanical interpretation.  In any event, the

Problem appears to be a mathematical problem worthy of further study.


REFERENCES


1.  Agmon, S., Spectral properties of Schrödinger operators and scattering theory, *Ann. Scuol. Norm. Sup. Pisa* 2 (1975), 151-218.
2.  Corbett, J. V. and Hurst, C. A., Are wave functions uniquely determined by their position and momentum distributions ?, *J. Austral.
    Math. Soc.* 20 (Ser. B) (1978), 182-201.
3.  Heisenberg, W., Die 'beobachtbaren Grössen' in der Theorie der
    Elementartcilchcn, I and II, *Z. Physik* 120 (1943), 513-538 and
    673-702.
4.  Kato, T., *Perturbation Theory for Linear Operators*, Springer, New
    York, 1966.
5.  Kato, T., Scattering theory, in *Studies in Applied Mathematics*, Vol. 7
    (ed. by A. H. Taub), Math. Assoc. Amer., Prentice-Hall, Englewood
    Cliffs, N. J.  (1971), 90-115.
6.  Kato, T. and Kuroda, S. T., Theory of simple scattering and eigenfunction expansions, in *Functional Analysis and Related Fields*
    (ed. by F. E. Browder), Springer, Berlin (1970), 99-131.
7.  Reed, M., and Simon, B., *Methods of Modern Mathematical Physics,
    Volume* I: *Functional Analysis*, Academic, New York, 1972.
8.  Taylor, J. R., *Scattering Theory*, John Wiley, New York, 1972.
9.  Vogt, A., Position and momentum distributions do not determine the
    quantum mechanical state, in *Mathematical Foundations of Quantum
    Theory* (ed. by A. R. Marlow), Academic, New York (1978), 365-372.
10. Wheeler, J. A., On the mathematical description of light nuclei by
    the method of resonating group structure, *Phys. Rev.* 52 (1937),
    1107-1122.
11. Wheeler, J. A., Semiclassical analysis illuminates the connection
    between potential and bound states in potential scattering, in
    *Studies in Mathematical Physics* (ed. by E. H. Lieb, B. Simon, and
    A. S. Wightman), Princeton U. Press, Princeton, N. J. (1976), 351-422.

# KALUZA AND KLEIN'S FIVE-DIMENSIONAL RELATIVITY

J. G. Miller

Department of Mathematics
Texas A&M University
College Station, Texas

The old unified field theory of Kaluza [1] and Klein [2] has been experiencing a revival since the development of gauge field theories in elementary particle physics. It is now recognized that Kaluza and Klein's 5-dimensional theory of relativity and the gauge field theories [3,4] can be formulated in the most elegant mathematical way in terms of connections in fibre bundles. The Kaluza-Klein theory is concerned with connections in principal fibre bundles and the gauge field theories with connections in vector bundles.

The Kaluza-Klein 5-dimensional theory of relativity is a classical theory that unifies the electromagnetic and gravitational fields in a special 5-dimensional metric and it is generally regarded as equivalent to the Einstein-Maxwell 4-dimensional theory of relativity. The equivalence of the 5-dimensional and 4-dimensional theories is actually only a local equivalence. This fact has not been brought out in the literature, even in the global formulations in which fibre bundles are employed. The curvature of a connection in a principal fibre bundle with structure group $S^1$ satisfies an integrality condition. Since the electromagnetic field in



Copyright © 1980 by Academic Press, Inc.
All rights of reproduction in any form reserved.
ISBN 0-12-473260-7

the 5-dimensional theory is identified with a multiple of this
curvature, it must satisfy a certain condition which is not
necessarily satisfied by the electromagnetic field in the 4-
dimensional theory.

For sake of simplicity, consider only the vacuum Einstein-
Maxwell equations

$$G = \kappa T, \quad \text{div} F = 0, \quad dF = 0, \tag{1}$$

where $G$ is the Einstein tensor and $T$ is the stress-energy
tensor of the electromagnetic field $F$ given by

$$T_{\mu\nu} = \frac{1}{4\pi}\left(F_\mu{}^\alpha F_{\nu\alpha} - \frac{1}{4} g_{\mu\nu} F_{\alpha\beta} F^{\alpha\beta}\right). \tag{2}$$

Greek indices run zero to three and Latin indices will run
from zero to four. The Einstein constant $\kappa = 8\pi G/c^4 = 8\pi$
in geometrized units in which Newton's constant $G$ and the
speed of light $c$ in vacuum are unity. The third equation
in (1) says that the electromagnetic field $F$ is a closed 2-
form. By the Poincaré lemma, a 1-form $A$ exists locally
such that $F = dA$. The first two equations in (1) can then
be derived from a variational principle by varying the functional

$$I_1 = \int \left( {}^{(4)}R - (\kappa/8\pi) F_{\alpha\beta} F^{\alpha\beta} \right) \left| \det \left( {}^{(4)}g_{\mu\nu} \right) \right|^{1/2} d^4x \tag{3}$$

with respect to the 4-dimensional metric ${}^{(4)}g_{\mu\nu}$ and the vector potential $A_\mu$. ${}^{(4)}R$ is the 4-dimensional scalar curvature.

The 5-dimensional spacetime manifold is a principal fibre
bundle [5] $P(M, S^1)$ with 4-dimensional base space $M$ and
structure group $S^1 = \{z \in \mathbb{C} \mid |z| = 1\}$. It was Klein's idea
to make the fibers compact in order to account for the quantization of charge [6]. This idea will be discussed later.
The mapping $\phi \mapsto e^{i\phi}$ defines an angular coordinate on $S^1$

which is periodic with period $2\pi$ and $\xi = d/d\phi$ is a left-invariant vector field on $S^1$. The Lie algebra of left-invariant vector fields on $S^1$ is isomorphic to the Lie algebra $\mathbb{R}$ of real numbers under the linear mapping that takes $\xi$ to the real number 1. From now on, the Lie algebra of $S^1$ will be identified with $\mathbb{R}$ under this isomorphism. This isomorphism is not unique since any nonzero left-invariant vector field on $S^1$ could be identified with 1.

Now let $\xi^*$ be the fundamental vector field [5] corresponding to $\xi$ which is tangent to the fibres of P. It is assumed that $\xi^*$ is a spacelike Killing and geodesic vector field with respect to the 5-dimensional Lorentz metric $^{(5)}g$ on P, which has signature $(-,+,+,+,+)$. As a result of these conditions on $\xi^*$, a 4-dimensional metric and electromagnetic field can be induced on M from $^{(5)}g$. Since $\xi^*$ is a Killing vector field, a connection 1-form $\omega$ in P can be defined by [7]

$$\omega(X) = {}^{(5)}g(\xi^*,X)/a^2, \tag{4}$$

where $a = (g(\xi^*,\xi^*))^{1/2}$ and $X$ is any tangent vector on P. Since $\xi^*$ is a geodesic vector field, $a$ is a constant. The positive number $a$ can be interpreted as the "radius" of $S^1$ since each fibre has circumference $2\pi a$.

The horizontal subspace $H_x$ of the tangent space $T_xP$ is defined to be the set of tangent vectors at $x$ that are annihilated by $\omega$. The vertical subspace $V_x$ is the 1-dimensional subspace of $T_xP$ tangent to the fibre at $x$. Note that $\xi^*$ is a vertical vector field with $\omega(\xi^*) = 1$. For any point $x$ in P, $T_xP = H_x \oplus V_x$. The induced 4-dimensional metric on M is the unique metric $^{(4)}g$ that satisfies [7]

$$^{(5)}g(hX,hY) = {}^{(4)}g(\pi_*X,\pi_*Y) \tag{5}$$

for arbitrary tangent vectors $X$ and $Y$ on $P$. The horizontal part of $X$ is denoted by $hX$ and the canonical projection of $P$ onto $M$ is denoted by $\pi$.

The curvature of $\omega$ is defined by $\Omega = D\omega$, where $D$ is exterior covariant differentiation [5]. Since $S^1$ is abelian $\Omega = d\omega$ and there exists a unique closed 2-form $f$ on $M$ such that $\Omega = \pi^*f$, where $\pi^*$ denotes the pullback map. The electromagnetic field $F$ is proportional to $f$ and the constant of proportionality is determined by examining the field equations for $^{(5)}g$.

The 5-dimensional metric $^{(5)}g$ can be expressed in terms of $^{(4)}g$ and $\omega$ by the equation

$$^{(5)}g(X,Y) = {}^{(4)}g(\pi_*X, {}_*Y) + a^2\omega \otimes \omega(X,Y), \tag{6}$$

where $X$ and $Y$ are tangent vectors on $P$. In a local coordinate system $(x^0,x^1,x^2,x^3)$ on $M$ and $x^4 = \phi$, $\omega_4 = 1$ and (6) becomes

$$^{(5)}g_{ij}dx^i dx^j = {}^{(4)}g_{\mu\nu}dx^\mu dx^\nu + a^2(d\phi + \omega_\mu dx^\mu)^2. \tag{7}$$

After a lengthy computation, the 5-dimensional scalar curvature $^{(5)}R$ is found to be related to the 4-dimensional scalar curvature $^{(4)}R$ and the curvature $\Omega$ by the formula

$$^{(5)}R = \pi^*({}^{(4)}R - \tfrac{1}{4} a^2 f_{\alpha\beta}f^{\alpha\beta}). \tag{8}$$

Consider the functional

$$\begin{aligned}
I_2 &= \int {}^{(5)}R|\det({}^{(5)}g_{ij})|^{1/2}d^5x \\
&= 2\pi a\int ({}^{(4)}R - \tfrac{1}{4} a^2 f_{\alpha\beta}f^{\alpha\beta})|\det({}^{(4)}g_{\mu\nu})|^{1/2}d^4x.
\end{aligned} \tag{9}$$

The variation of $I_2$ with respect to the special 5-dimensional metrics of the form (6) for fixed $a$ yields the

5-dimensional vacuum field equations.  Now it is clear that the electromagnetic field  $F$  should be defined by

$$F = a(2\pi/\kappa)^{1/2} f \tag{10}$$

so that the following theorem holds.

THEOREM 1.  If  $(P, {}^{(5)}g)$  is a solution of the vacuum 5-dimensional field equations, then the induced 4-dimensional Lorentz metric  ${}^{(4)}g$  on  $M$  defined by (5) and the 2-form  $F$  defined by (10) satisfy the vacuum Einstein-Maxwell equations (1).

The curvature of an  $S^1$  bundle satisfies an integrality condition, which imposes a condition on the electromagnetic field because of (10).  Let  $\mathbb{Z}$  be the set of integers and let  ${}_\infty H_2(M,\mathbb{Z})$  be the second  $C^\infty$  singular homology group of  $M$  with integer coefficients.  Then the integrality condition on the curvature can be stated as follows [8].

THEOREM 2.  If  $\Omega = \pi^* f$  is the curvature of a connection  ω in a principal fibre bundle  $P(M,S^1)$ , then  $\frac{1}{2\pi} \int_z f \in \mathbb{Z}$  for every 2-cycle  $z \in {}_\infty H_2(M,\mathbb{Z})$ .

The factor of  $2\pi$  is due to the choice of identification of the Lie algebra of  $S^1$  with  $\mathbb{R}$ .  Theorem 2 implies that the electromagnetic field induced from a 5-dimensional metric by (10) must satisfy

$$\frac{1}{2\pi a} \int_z (\kappa/2\pi)^{1/2} F \in \mathbb{Z} \tag{11}$$

for every 2-cycle  $z \in {}_\infty H_2(M,\mathbb{Z})$ .  The converse of Theorem 2 is known as Kostant's theorem [8] in geometric quantization. (It is usually stated for the associated complex line bundles rather than the principal bundles.)

THEOREM 3.   (Kostant) If  f  is a closed 2-form on  M  such that $\frac{1}{2\pi}\int_z f \in \mathbb{Z}$  for every 2-cycle  $z \in {}_\infty H_2(M,\mathbb{Z})$, then there exists a principal fibre bundle  $P(M,S^1)$  with connection  $\omega$  whose curvature  $\Omega = \pi^* f$.  If  M  is simply connected, the bundle  P  is unique up to bundle isomorphisms.

Given a 4-dimensional metric  ${}^{(4)}g$  and electromagnetic field  F  that satisfies (11) for some positive real number a, Kostant's theorem provides the existence of a principal fibre bundle and connection for defining the 5-dimensional metric  ${}^{(5)}g$  by (6).

THEOREM 4.  If  $(M, {}^{(4)}g, F)$  is a solution of the Einstein-Maxwell vacuum equations and if there exists a positive real number  a  such that (11) holds, then there exists a principal fibre bundle  $P(M,S^1)$  with connection  $\omega$  such that the 5-dimensional metric  ${}^{(5)}g$  defined on  P  by (6) satisfies the 5-dimensional vacuum equations and the 4-dimensional metric and electromagnetic field induced on  M  from  ${}^{(5)}g$  are the given  ${}^{(4)}g$  and  F, respectively.

Thus condition (11) is both a necessary and sufficient condition on the electromagnetic field for the 5-dimensional metric to exist globally.  If the electromagnetic field is exact, then condition (11) is satisfied trivially for every positive real number  a  and the principal fibre bundle  P  is trivially  $M \times S^1$.  In this case, the radius of  $S^1$  can be arbitrarily chosen.  If the second Betti number of  M  is one and  F  is not exact, then there exists a discrete set of positive real numbers that satisfy condition (11) and the corresponding fibre bundles are nontrivial.  If the second

Betti number of  M  is greater than one and  F  is not exact,
then there may not exist a positive real number  a  such that
(11) holds.  Thus a globally defined 5-dimensional metric
corresponding to a given 4-dimensional metric and electromagnetic field may not exist.

Some examples will illustrate these remarks.  The spherically symmetric, magnetically charged black hole of mass  M
and magnetic charge  g  is given by the Reissner-Nordstrøm
metric

$$ds^2 = -\Delta r^{-2}dt^2 + \Delta^{-1}r^2dr^2 + r^2(d\theta^2 + \sin^2\theta d\varphi^2), \qquad (12)$$

where

$$\Delta \equiv r^2 - 2M(G/c^2)r - g^2G/c^4 \qquad (13)$$

and

$$F = g \sin\theta d\theta \wedge d\varphi = \frac{g}{r^2} (rd\theta) \wedge (r \sin\theta d\varphi). \qquad (14)$$

The topology of  M  is  $\mathbb{R}^2 \times S^2$  and its second Betti number
is one.  F  is not exact and condition (11) becomes

$$\frac{1}{2\pi a} \int_{S^2} (\kappa/2\pi)^{1/2}F = 2g(\kappa/2\pi)^{1/2}/a = m \qquad (15)$$

for some integer  m.  The principal fibre bundles  $P_m$  are
given by

$$P_m = \mathbb{R}^2 \times (m,1), \qquad (16)$$

where  (m,1)  is a lense space [9], and the curvature
$\Omega_m = (m/2g) \pi_m^*F$.  Thus for a given value of  g, there are an
infinite number of distinct nontrivial fibre bundles  $P_m$  and
5-dimensional metrics  $^{(5)}g_m$  with fibers of radius
$a = 2g(\kappa/2\pi)^{1/2}/m$  which induce the 4-dimensional metric (12)
and the electromagnetic field (14).  Kaluza and Klein's 5-di-
mensional relativity is a classical theory and there is no

restriction on the value of the magnetic charge  $g$  since there is no a priori value for the radius  $a$  of the fibers. The charge quantization condition arises only when quantum mechanics is considered.

As another example, consider  $M = \mathbb{R} \times T^3$ .  Let  $t$  be the natural coordinate on  $\mathbb{R}$  and let  $(\theta^1, \theta^2, \theta^3)$  be periodic co-ordinates on the 3-dimensional torus  $T^3$ , each with period  $2\pi$ .  If  $(M, {}^{(4)}g, F)$  is a solution of the Einstein-Maxwell equations, then  $F$  is closed.  Since the second Betti number of  $M$  is three and  $\{d\theta^2 \wedge d\theta^3,\ d\theta^3 \wedge d\theta^1,\ d\theta^1 \wedge d\theta^2\}$  is a basis of  $H^2(M)$ , the second de Rham cohomology group of  $M$ , the closed 2-form  $F$  can be written as

$$F = \beta_1 d\theta^2 \wedge d\theta^3 + \beta_2 d\theta^3 \wedge d\theta^1 + \beta_3 d\theta^1 \wedge d\theta^2 + \tilde{F}, \tag{17}$$

where  $\beta_1, \beta_2$  and  $\beta_3$  are real numbers and  $\tilde{F}$  is an exact 2-form.  The condition (11) is satisfied for some positive real number  $a$  if and only if  $\beta_2/\beta_1$  and  $\beta_3/\beta_1$  are rational numbers, provided  $\beta_1 \neq 0$ .  If this condition is not met, a global 5-dimensional metric that induces  ${}^{(4)}g$  and  $F$  will not exist.  Condition (11) is a rather severe restriction on the values of the real numbers  $\beta_1, \beta_2$  and  $\beta_3$ .

So far no quantum mechanics has entered the discussion. Klein [6] imposed the condition of compactness on the fibers in order to account for the quantum nature of charge, which arises when quantization of the 5-dimensional geodesic equations is considered.  First, consider the trivial case  $P = M \times S^1$ .  The 5-dimensional geodesic equations on  $P$  reduce to the Lorentz force law equations on  $M$  for a charged particle with charge  $e$  moving in the electromagnetic field  $F$  if the conserved momentum  $p_\phi$  is given by

$$p_\phi = (ea/c)(2\pi/\kappa)^{1/2}. \tag{18}$$

In quantum mechanics, the operator corresponding to $p_\phi$ is $(\hbar/i)\xi^* = (\hbar/i)\partial/\partial\phi$ and its eigenvalues are $\mu = n\hbar$, where $n$ is an integer. The quantization condition on the electric charge $e$ is

$$(ea/c)(2\pi/\kappa)^{1/2} = n\hbar. \tag{19}$$

Given the experimental value of the elementary quantum of charge, the radius $a$ of the fibers is determined by setting $n = 1$ in (19). The result is

$$a = 2e/\alpha \cong 3.78 \times 10^{-32} cm, \tag{20}$$

where $e = eG^{1/2}/c^2$ is the elementary quantum of charge in geometrized units and $\alpha$ is the fine structure constant. This small value of $a$ would explain the lack of evidence for the fifth dimension if it exists.

The quantization condition (19) can also be derived by minimal coupling of the $U(1)$ gauge field and the charged particle. Let $\rho_n$ be the unitary representation of $S^1$ on $\mathbb{C}$ defined by $\rho_n(e^{i\phi})(z) = e^{-in\phi} \cdot z$, where $n$ is a nonzero integer. Let $L_n$ be the complex line bundle [5] associated to $P = M \times S^1$ and the representation $\rho_n$. The curvature of $L_n$ is the 2-form $R_n$ on $M$ defined by

$$[\nabla_X, \nabla_Y]\sigma - \nabla_{[X,Y]}\sigma = iR_n(X,Y)\sigma, \tag{21}$$

where $X$ and $Y$ are vector fields on $M$, $\sigma$ is a section in $L_n$ and $\nabla$ is the covariant differentiation in $L_n$ obtained by parallel translation with respect to the connection $\omega$ in $P$. The curvatures $\Omega$ and $R_n$ are related by

$$R_n = -nf. \tag{22}$$

Expressed in local coordinates, minimal coupling is the condition that covariant differentiation should be given by $\nabla_\mu = \partial_\mu - (ie/\hbar c)A_\mu$, where $F = dA$. By the definition of the curvature of $L_n$, minimal coupling requires that

$$R_n = -(e/\hbar c)F. \tag{23}$$

The quantization condition (19) follows from (10), (22) and (23).

Now consider the first example again. For each $m$, the geodesic equations on $P_m$ reduce to the Lorentz force law equations on $M$ for a charged particle with charge $e$ moving in the electromagnetic field $F$ in (14) if the conserved momentum $p_\phi$ is given by (18). The quantum operator corresponding to $p_\phi$ is $(\hbar/i)\xi_m^* = (\hbar/i)\partial/\partial\phi$ and its eigenvalues are $\mu = n\hbar$, where $n$ is an integer. The quantization condition on the electric charge is again (19) and it is independent of $m$. The radius $a$ of the fibers is independent of $m$ and is given by (20). The quantization condition on the magnetic charge

$$eg/\hbar c = \tfrac{1}{2}mn \tag{24}$$

is obtained by substituting (15) into (19). For $m = 1$, this is Dirac's quantization condition [10] and for $m = 2$, Schwinger's quantization condition [11]. The quantization condition on the magnetic charge depends on the choice of the 5-dimensional classical configuration space $P_m$ and this choice gives a geometric explanation for the different quantization conditions. This point has been made before [12].

The quantization condition on the magnetic charge can also be derived by minimal coupling [13]. Let $L_{mn}$ be the complex line bundle associated to $P_m$ and the representation $\rho_n$ and let $R_{mn}$ be the curvature of $L_{mn}$. Then

$$R_{mn} = -nf_m \qquad (25)$$

and by (10) and (15)

$$\frac{1}{2\pi} \int_{S^2} f_m = \frac{1}{2\pi a} \int (\kappa/2\pi)^{1/2} F = m. \qquad (26)$$

From these equations $\frac{1}{2\pi} \int_{S^2} R_{mn} = -mn$ and from (14) $\frac{1}{2\pi} \int_{S^2} F = 2g$. The condition of minimal coupling $R_{mn} = -(e/\hbar c) F$

now implies the quantization condition (24).

The set of complex valued equivariant functions with respect to $\rho_n$ on $P_m$ is equal to the set of eigenfunctions of $(\hbar/i) \xi_m^*$ with eigenvalue $n\hbar$ and this set is in one-to-one correspondence with the set of sections of $L_{mn}$. In a local coordinate system $\psi(x^0, x^1, x^2, x^3, \phi) = \tilde{\psi}(x^0, x^1, x^2, x^3) e^{in\phi}$, where $\psi$ is the eigenfunction and $\tilde{\psi}$ the section. Either an eigenfunction or the corresponding section can be used to describe the quantum mechanical state of the charged particle. The wave function $\psi$ satisfies the 5-dimensional Klein-Gordon equation on $(P_m, {}^{(5)}g_m)$ and a superselection rule (so that wave functions for different values of $n$ are not super-imposed) and the corresponding wave section $\tilde{\psi}$ satisfies the 4-dimensional Klein-Gordon equation minimally coupled to the electromagnetic field $F$ in (14) by means of covariant differentiation in $L_{mn}$ [14].

REFERENCES

1. Th. Kaluza, Sitz. Preus. Akad. Wiss. 966 (1921).
2. O. Klein, Z. Phys. 37, 895 (1926).
3. C. N. Yang and R. L. Mills, Phys. Rev. 96, 191 (1954).
4. R. Utiyama, Phys. Rev. 101, 1597 (1956).

5.  S. Kobayashi and K. Nomizu, "Foundations of Differential Geometry" (Interscience, New York, 1963), Vol. 1.
6.  O. Klein, *Nature* *118*, 516 (1926).
7.  J. Sniatycki and W. M. Tulczyjew, *Ann. Inst. Poincaré* *15*, 177 (1971).
8.  D. J. Sims and N. M. J. Woodhouse, "Lectures on Geometric Quantization" (Lecture Notes in Physics, Vol. 53, Springer-Verlag, New York, 1976).
9.  N. Steenrod, "The Topology of Fibre Bundles" (Princeton U.P., Princeton, N. J., 1951), pp. 135.
10. P. A. M. Dirac, *Proc. Roy. Soc. (London)* *A 133*, 60 (1931).
11. J. Schwinger, *Phys. Rev.* *144*, 1087 (1966); *Science 165*, 757 (1969).
12. J. G. Miller, *J. Math. Phys.* *17*, 643 (1976).
13. W. Greub and H.-R. Petry, *J. Math. Phys.* *16*, 1347 (1975).
14. J.-M. Souriau, *Nuovo Cimento 30*, 565 (1963).

# MODERN MATHEMATICAL TECHNIQUES

# IN THEORETICAL PHYSICS

Carroll F. Blakemore

Department of Mathematics
University of New Orleans
New Orleans, Louisiana

## I. INTRODUCTION

In the last few years, it has become evident that the unfortunate isolation of the theoretical physicist from much of modern mathematics as well as that of the mathematician from exciting developments in physics and astrophysics is beginning to disappear.  This is vividly illustrated by the number of conferences [13], [18], [22], [24], [70], to mention only a few, which have brought together both physicists and mathematicians during the past few years.  The Loyola Conference on Quantum Theory and Gravitation was typical in this respect.  Now it is becoming the rule rather than the exception to find both physicists and mathematicians on somewhat of an equal footing in discussing such topics as differentiable manifolds, differential forms, fibre bundles, and connections as well as gauge fields, general relativity, and black holes.  Certainly, with few exceptions, physicists and mathematicians are in agreement that such communication will benefit both disciplines.

At the time of the Loyola conference, there is great

233

Copyright © 1980 by Academic Press, Inc.
All rights of reproduction in any form reserved.
ISBN 0-12-473260-7

excitement in the world of theoretical physics concerning the possible unification of the four fundamental interactions - the strong, the electromagnetic, the weak, and the gravitational - under one "grand unified theory."  The Weinberg - Salam unified gauge theory of the weak and electromagnetic interactions seems quite promising.  Also, there is great hope for supersymmetry and local supersymmetry = supergravity.  Finally, a theory of "Quantum Gravity" is being actively sought by many investigators.  Needless to say, these are all extraordinary undertakings but, even though progress has been rapid in some respects, there are numerous difficulties confronting these programs.

At the Loyola conference, which was concerned with quantum theory and gravitation, several participants (Brans, Finkelstein, Marlow, Wheeler) voiced the opinion that it was probably a mistake to rely on the notion of a differentiable manifold as being fundamental for a model of space-time. They warned against the pitfalls of pushing the "differential geometry" approach too far when it comes to the quantum level. However, in what follows, I wish to illustrate how several recent ideas have been enriched and some old ones clarified by the techniques of modern differential geometry.  In what follows, I will select illustrations of this point of view from each of classical mechanics, classical electrodynamics, general relativity theory, elementary particle physics, and gauge field theory.  Also, I will mention some mathematical techniques used in quantum mechanics.  I can think of no better way to begin than with a timely quotation from an article by Saunders MacLane [69] where the two parenthetical expressions are my own:

"The thesis of this article is that classical mechanics (parts of theoretical physics) and recent conceptual methods in abstract mathematics have a lot to do with each other.  Many abstract ideas of pure geometry originally arose in the study of mechanics; for example, the cotangent bundle of a differentiable manifold appeared first as the phase space of Hamiltonian mechanics.  Many cumbersome developments in the standard treatments of mechanics (parts of theoretical physics) can be simplified and better understood when formulated with modern conceptual tools, as in the well-known case of the "universal" definition of tensor products of vector spaces to simplify some of the notational excesses of tensor analysis as traditionally used in relativity theory."

## II. CLASSICAL MECHANICS [2, 4, 68, 94]

In what follows we begin with Newtonian mechanics and proceed to the Lagrangian and Hamiltonian formulations of classical mechanics.  (Here our presentation follows that of V. I. Arnold [4].)  Finally, we present a brief discussion of the Hamilton-Jacobi equation following MacLane [69].

In the Newtonian case we are interested in the motion of a system of n point masses in Euclidean three space $R^3$.  The laws of Newtonian mechanics are invariant under the six dimensional group E(3) of Euclidean motions.  (We recall that the Euclidean group E(3) is topologically  $R^3 \times 0(3)$  where $0(3)$ is the orthogonal group in dimension three and that algebraically E(3) is the semidirect product of $R^3$ by $0(3)$.  Finally, E(3) is a Lie transformation group and $R^3$ is realized as a homogeneous coset space $E(3)/O(3)$ determined by the action of $E(3)$ on $R^3$.)  Of course, Newtonian mechanics is also invariant under the larger group of Galilean transformations of space and time.  The laws of conservation of a Newtonian system are determined by the Euclidean motions which leave the potential energy of the system invariant.  Many important problems from classical mechanics can be solved using the Newtonian formalism.  Among them is the problem of motion in a central force field which is of fundamental importance in celestial mechanics.

The Lagrangian formalism is concerned with describing motion in a mechanical system in terms of the configuration space.  If there are no constraints, then the configuration space of a system of n point masses is just the direct product of n copies of $R^3$.  However, when constraints are imposed the configuration space is a differentiable manifold on which its group of diffeomorphisms acts.  As in the Newtonian case, the laws of Lagrangian mechanics are invariant under this group.  Mathematically speaking, a Lagrangian mechanical system is prescribed by a differentiable manifold = the configuration space and a smooth real-valued function on its tangent bundle = the Lagrangian.  Every one-parameter group of diffeomorphisms of the configuration space which leaves the Lagrangian invariant determines a conservation law.  In a Newtonian potential system the Lagrangian is just the kinetic energy minus the potential energy.  The Lagrangian formalism is particularly suited for handling problems in the theory of small oscillations and in studying the dynamics of a rigid body.

Just as Lagrangian mechanics is concerned with configuration space, the Hamiltonian formalism is concerned with phase

space.  The phase space of Hamiltonian mechanics has the
structure of a symplectic manifold.  (We recall that a
symplectic manifold is an ordered pair $(M,\omega)$ where M is a
smooth (paracompact Hausdorff) manifold and $\omega$ is a smooth nondegenerate closed 2-form on M.  It follows from Darboux's theorem that M must be even dimensional.)  The laws of Hamiltonian mechanics are invariant under the group of symplectic diffeomorphisms which act on $(M,\omega)$.  Mathematically speaking, a
Hamiltonian mechanical system is prescribed by an even dimensional smooth (paracompact Hausdorff) manifold M = the phase
space, a symplectic structure $\omega$ on M = the Poincaré integral
invariant, and a smooth real-valued function on M = the Hamiltonian.  Every one-parameter group of symplectic diffeomorphisms of the phase space which leaves the Hamiltonian invariant determines a conservation law.  Hamiltonian mechanics contains Lagrangian mechanics as a special case.  In this case
the phase space is the cotangent bundle of the configuration
space and the Hamiltonian is the Legendre transform of the
Lagrangian.  The Hamiltonian formalism allows us to solve such
problems as "attraction by two stationary centers" and "determining geodesics on the triaxial ellipsoid" which do not yield
solutions by other means.  Also, the Hamiltonian formalism is
extremely useful for the approximate methods of perturbation
theory.  It is also useful in the analysis of motion in complicated systems as in ergodic theory and statistical mechanics.  Finally, the Hamiltonian formalism has a connection with
geometric optics and quantum mechanics.  For some interesting
work concerning the geometric quantization of classical systems as developed by B. Kostant [63] and J. M. Souriau [92],
we refer the reader to Simms and Woodhouse [91] and Śniatycki

[93] and the references there.

Enjoying a particularly close relationship with quantum mechanics is Hamilton-Jacobi theory.  The Hamilton-Jacobi equation can be given a simple interpretation using the language of differentiable manifolds.  Let $H: T^*(M) \longrightarrow R$ be the Hamiltonian where $T^*(M)$ denotes the cotangent bundle of configuration space M.  If $S: M \longrightarrow R$ is a smooth real-valued function, then $dS: M \longrightarrow T^*(M)$ is a smooth 1-form, i.e., a cross-section of the cotangent bundle where d denotes exterior differentiation.  The composition of H and dS, denoted HodS, is a smooth real-valued function on M.  If we let HodS be equal to zero, then we obtain an equation HodS = 0 on M called the Hamilton-Jacobi partial differential equation.  If we write the Hamiltonian H as a function $H(q^1, \ldots, q^n, p_1, \ldots, p_n)$ of local coordinates and the differential $dS: M \longrightarrow T^*(M)$ as the map $m \longmapsto d_m S$ where the point m has the coordinates $q^1, \ldots, q^n$ and $\frac{\partial}{\partial q^1}, \ldots, \frac{\partial}{\partial q^n}$ determine a basis for the tangent space to M at the point m, then the Hamilton-Jacobi equation appears in its usual form as $H(q^1, \ldots, q^n, \frac{\partial S}{\partial q^1}, \ldots, \frac{\partial S}{\partial q^n}) = 0$.  It is a partial differential equation of the first order in S.

The methods of modern differential geometry can do much to clarify many concepts in classical mechanics which are usually left vague and almost devoid of mathematical rigor.  (Perhaps as a Herculean exercise one should rewrite Goldstein [41] using the language of modern differential geometry.)  We refer the reader to Abraham and Marsden [2], Arnold [4], and MacLane [68] for a detailed development of this approach.  Also, see the interesting paper by Arthur Komar [62] containing his constraint formalism for classical mechanics and the references there.

   III. CLASSICAL ELECTRODYNAMICS [33, 34, 35, 74, 103]

   It is well-known that Maxwell's equations can be written
in a very concise and elegant way by using E. Cartan's beautiful theory of exterior differential forms.  In particular,
Maxwell's source-free equations $\nabla \cdot \vec{B} = 0$ and Faraday's law:
$\nabla \times \vec{E} = - \frac{\partial \vec{B}}{\partial t}$ are equivalent to the equation dF = 0 where F is
the electromagnetic field 2-form (or Faraday) and d is the ex-
terior derivative.  Finally, Gauss' law: $\nabla \cdot \vec{D} = \rho$ and the
Ampère-Maxwell law: $\nabla \times \vec{H} = \vec{J} + \frac{\partial \vec{D}}{\partial t}$ are equivalent to the equation d * F = J where * is the Hodge star operator and J is the
current 3-form.  (Of course, the domain of definition of all
our forms is Minkowski space-time which is a pseudo-Riemannian
manifold of dimension four; hence, the Hodge star operator
maps p-forms to (4-p)-forms.  Thus * F is a 2-form and d * F
is a 3-form.)

   Perhaps not so well-known is a remark of John Wheeler
([110], p. 84) to the effect that, in a certain sense,
Faraday's law is a consequence of $\nabla \cdot \vec{B} = 0$ and the Ampère-
Maxwell law is a consequence of Gauss' law.  The precise details have been worked out by Theodore Frankel [34], [35] in
Minkowski space-time using the formalism of exterior differential forms.  Frankel proves that if, in all inertial systems,
one has $\nabla \cdot \vec{B} = 0$, then dF = 0 and consequently each inertial
observer also finds that $\nabla \times \vec{E} = - \frac{\partial \vec{B}}{\partial t}$ .  Also, he proves that
if, in all inertial systems, one has $\nabla \cdot \vec{D} = \rho$, then d * F = J
and consequently each inertial observer also finds that
$\nabla \times \vec{H} = \vec{J} + \frac{\partial \vec{D}}{\partial t}$ .  This is indeed satisfying since $\nabla \cdot \vec{B} = 0$
and Gauss' law are relatively simple whereas Faraday's law and

the Ampère-Maxwell law are more complicated.

It is unfortunate that the equation of motion for a particle in an electromagnetic field, i.e., the Heaviside-Lorentz
force law $\vec{F} = q[\vec{E} + (\vec{v} \times \vec{B})]$ does not follow from Maxwell's
equations.  This stands in vivid contradistinction to the gen-
eral theory of relativity where the geodesic equations of motion follow from the Einstein field equations [30].  This is
intimately associated with the fact that Maxwell's equations
are linear whereas the Einstein equations are nonlinear.  Certainly, as the late Werner Heisenberg advocated [51], there is
much to be gleaned from physical laws expressed by nonlinear
equations.  Here one is reminded of a "missed opportunity" in
the sense of Freeman Dyson [29].  Only recently (solitons, the
Korteweg-deVries equation, etc. [12]) have physicists and
mathematicians combined to mount a serious attack on the fundamental, albeit difficult, problems associated with nonlinear
phenomena.

IV. GENERAL RELATIVITY THEORY [49, 74, 88, 97]

Einstein's general theory of relativity published in 1916
is surely one of the most magnificent creations of the human
mind during this century.  It is both a theory of gravitation
and a theory of space-time geometry.  Einstein's deep insight
led him to identify the gravitational field with the space-
time geometry and his theory is the most viable theory of
gravitation that we have after over sixty years of observational and experimental tests.

Differential geometry has played an important part in the
mathematical foundations of general relativity since its in-
ception.  In orthodox general relativity, one thinks of space-

time as a certain connected four-dimensional pseudo-Riemannian

manifold with a Lorentz metric.  In the next paragraph let us

devote a few lines to making this quite precise.

A pseudo-Riemannian manifold is an ordered pair $(M,g)$

where M is a smooth manifold (Hausdorff, but not necessarily

paracompact) and g is a function which assigns to each point

$p \varepsilon M$ a nondegenerate symmetric bilinear form $g_p$ on the tan-

gent space to M at p, denoted $M_p$, with g smooth, i.e., for any

two smooth vector fields X, Y on M, the function $g(X,Y): M \rightarrow R$

defined by $[g(X,Y)](p) = g_p(X_p,Y_p)$ for each $p \varepsilon M$ is smooth.

(If $g_p$ is positive-definite on $M_p$ for each $p \varepsilon M$, then $(M,g)$

is called a Riemannian manifold.  Let me state here that I

think mathematicians and physicists should always state wheth-

er their results do or do not hold for pseudo-Riemannian man-

ifolds and, in case they do not, provide appropriate counter-

examples exhibiting why they must be restricted to Riemannian

manifolds.  There is a big difference between the indefinite

and the positive-definite cases.)  A space-time is then a con-

nected four-dimensional pseudo-Riemannian manifold $(M,g)$ of

Lorentz signature $(+ - - -)$.  By Lorentzian we mean that for

each $p \varepsilon M$ there exists a basis in $M_p$ relative to which the

matrix of $g_p$ has the form Diag $(1, -1, -1, -1)$.  (Thus, each

$M_p$ is a Minkowski space-time which is "locally" what one would

expect.)  One then says that g is a Lorentz metric.  It fol-

lows from a result of Geroch [38] that M is also paracompact.

Hence, there is no loss of generality in assuming that M is

paracompact from the beginning.  Since M is connected, para-

compactness is equivalent to second countability.  (Many au-

thors [88] also require that $(M,g)$ be oriented and time-

oriented (they are independent concepts) and that it carry the unique Levi-Civita connection D determined by g.)

Now that we know what we mean by a space-time as defined in the previous paragraph, an obvious question arises:  Suppose that M is a connected paracompact Hausdorff four-dimensional smooth manifold.  When does M admit a Lorentz metric?  The answer is exactly when M admits a smooth nonvanishing vector field [82] or equivalently (provided M is compact) that the Euler characteristic of M be zero.  Techniques from differential geometry show that this is always the case if M is not compact.  (The case for M compact is physically not very interesting because of causality violations such as the existence of closed timelike curves [49].  Even so, a manifold such as the four-dimensional sphere $S^4$ would be eliminated as a candidate for a space-time since its Euler characteristic is two.)  We note that at the Loyola conference Maurice Dupré suggested some interesting ways in which surgery theory and cobordism theory might be employed in studying space-time.  We refer the reader to his paper in these proceedings.  Also, see Dupré [28] and Yodzis [115].

For a given space-time (M,g) with Levi-Civita connection D determined by g, one can write down the Einstein equation for matter fields in coordinate-free language as Ric $- \frac{1}{2}Sg + \Lambda g =$ $-\kappa T$ where Ric is the Ricci tensor, S is the Ricci scalar curvature, $\Lambda$ is the cosmological constant, T is the energy-momentum tensor and $\kappa = \frac{8\pi G}{c^4}$ where G is the Newtonian gravitational constant and c is the speed of light.  In local coordinates one has the more familiar $R_{\mu\nu} - \frac{1}{2}g_{\mu\nu}R + \Lambda g_{\mu\nu} = -\kappa T_{\mu\nu}$ with R denoting the scalar curvature.  These form a set of ten nonlinear partial differential equations of the second-order.  We make no effort to "derive" the Einstein field equations, but refer the reader to any number of the standard works [74], [95], [97] on general relativity where some type of development of the field equations is given. (For a novel development which differs from Einstein's original approach, we refer the reader to Bondi in [16].) I have never been happy with any of the developments which I have seen and look forward to a more mathematically rigorous derivation of the Einstein field equations from some physically reasonable set of postulates. (However, see section 17.5 of [74] in this regard.)

Also, I am perplexed by the widely divergent viewpoints of some authorities concerning the proper position of the so-called "Principle of Equivalence" in the edifice that is the general theory of relativity. For instance some authorities such as S. Weinberg ([105], p. 289) consider it to be a foundation stone in the theory whereas others such as J. L. Synge ([95], pp. ix-x) seem to think it is only of historical interest and has no place in the theory today. This is a confusing state of affairs and a careful "axiomatic development" of general relativity should give a mathematically meaningful statement of a principle of equivalence (whether it be a "weak" form or a "strong" form or some other modification). James Anderson addresses himself to this problem in [3] (also see his reference 2 there), but I am still not convinced by his arguments. He takes the point of view that fundamental to general relativity is the principle that any inertial effect can be duplicated by some gravitational effect. However, the converse need not hold. Finally, just what is the principle of equivalence? Hans Ohanian attempts to answer this question in [79] where he discusses the weak and strong forms of the equivalence principle. He argues that the strong principle of equivalence is false and presents several examples involving tidal effects to support his claim. Also, he presents a somewhat more mathematical formulation of the strong principle. For a rebuttal to Ohanian's arguments, we refer the reader to Allan Walstad in [104]. I look forward to a resolution of the unfortunate state of affairs surrounding the "equivalence principle" and its "many faces" at some time in the not too distant future.

As Hawking and Ellis [49] have pointed out, Einstein's general theory of relativity leads to two remarkable predictions about our universe: The first is that stars of sufficiently high mass will undergo uninhibited gravitational collapse to form a black hole containing a singularity. The second is that there is a singularity in the past history of the universe which could be identified with the creation of our universe. These are interpretations of the intriguing singularity theorems of Hawking and Penrose [50] and their derivation depends on very reasonable physical and mathematical assumptions. Fundamental to their development of these results is the causal structure of space-time. There is much beautiful mathematics used here. For excellent treatments, the reader is referred to Penrose [82] and Hawking and Ellis [49]. For a presentation of Schmidt's b-boundary (a construction which attaches to any space-time a boundary which can be used to define singularities) and its modifications by Clarke, the reader is referred to C. J. S. Clarke and B. G. Schmidt in [20] for "the state of the art." Finally, for a development of the requisite material from modern differential geometry as well as a reasonably self-contained discussion of geometrical singularities in space-time, we strongly recommend C. T. J.

Dodson [26].

In conclusion, we mention the very important work of Stephen Hawking [47] published in 1975 concerning the quantum evaporation of black holes and particle creation by black holes.  Here, Hawking treats the space-time metric g classically, but treats the matter fields quantum mechanically. Hawking has presented us with a beautiful synthesis of concepts from general relativity, quantum mechanics, and thermodynamics.  A classical black hole can absorb, but not emit particles.  However, quantum mechanical effects permit black holes to create and emit particles as if they were hot bodies with temperature approximately $10^{-6} (\frac{M_\odot}{M})\,^\circ K$ where $M_\odot$ is the solar mass in grams and $M$ is the mass in grams of the black hole.  Thus, for solar mass black holes, this temperature would be negligible.  However, for a mini black hole (provided such a creature exists) with mass approximately $10^{15}$ grams the temperature would be roughly $10^{12}$ degrees Kelvin.  It is hoped that the Hawking process is an approximation to some deeper theory of quantum gravity in which space-time is quantized. Hopes and aspirations are high and perhaps Hawking's work is the first step on the road to a fullblown theory of quantum gravity.  (For more results on the production of particles by strong gravitational fields we refer the reader to the article by Leonard Parker in [81] and the references there.)


## V. QUANTUM MECHANICS [23, 54, 57, 66, 85]

Everyone is aware of the important role played by mathematics in the foundations of quantum mechanics.  From its beginnings with Heisenberg's "matrix mechanics" in 1925 and Schrödinger's "wave mechanics" in 1926, operator-theoretic

concepts have played an important part in the development of
quantum mechanics.  As early as 1932, with John von Neumann's
<u>Mathematical</u> <u>Foundations</u> <u>of</u> <u>Quantum</u> <u>Mechanics</u> [102], the use
of operators on Hilbert space had reached a fairly sophisti-
cated level.  In what follows, we will very briefly mention a
few areas of mathematics which have proved useful in the
mathematical foundations of quantum mechanics.

In 1936, G. Birkhoff and J. von Neumann [11] introduced
the formalism of lattice theory to study the logic of quantum
mechanics.  This was an outgrowth of von Neumann's Hilbert
space model of quantum mechanics which contained the basic
concepts for the lattice theoretical approach to the foundations of quantum mechanics.  Two decades later, G. W. Mackey
[65] had formulated an axiomatic approach to quantum mechanics
which was very attractive.  The highly influential <u>Introduc-</u>
<u>tion</u> <u>to</u> <u>Hilbert</u> <u>Space</u> by Paul Halmos [45] had appeared in
1951.  It contained the lattice theoretical properties of
closed subspaces and projection operators.  In 1957, Andrew
Gleason [40] published a very important paper containing what
has come to be known as "Gleason's theorem" which pointed the
way to a respectable theory of measures on nondistributive
lattices.  (We were indeed fortunate to have both Gleason and
Halmos in attendance at the Loyola conference.)  The state of
the art as of roughly 1963 is contained in Mackey's influential book [66].  In the next few years, such expressions as
"orthomodular ortholattice" and "Baer$^{*}$ - Semigroup" became
commonplace.  Contributions to these and other areas were made
by several workers in the field such as the valuable work of
Dick Greechie [42] (who was also at the Loyola conference),
the late J. M. Jauch [57], B. Mielnik [73], C. Piron [84], and

V. S. Varadarajan [100], to mention only a few. For developments in the last decade, we refer the reader to Hooker [54], Marlow [70], Mittelstaedt [75], Piron [85] and the references there.

Of course, the lattice-theoretic approach to the foundations of quantum mechanics is not the only one. The algebraic approach to the foundations of quantum mechanics via $C^*$ - algebras initiated by Irving E. Segal [89] in 1947 has been of considerable importance. We mention the fundamental work of Haag and Kastler [44] in this regard. Also, we recommend [58] for a detailed presentation of the $C^*$ - algebra approach. As far as axiomatic quantum field theory is concerned one has the important contribution of Bogolubov, Logunov, and Todorov [14], better known as "BLT" by practitioners of the art. (Steve Fulling, who attended the Loyola conference, was in charge of translating BLT from the original Russian.)

Needless to say, there is much important mathematics being used in probing the foundations of quantum mechanics - operator theory [86], harmonic analysis [43], [67], the Schwartz theory of distributions [14], [19], several complex variables (e.g., the edge-of-the-wedge theorem) [101], symplectic geometry [71], etc. (We have only touched on a very small part of the ongoing work.) As Jerry Goldstein pointed out at the Loyola conference, there is a mathematical theory of scattering theory which had its origins in scattering problems in particle physics and with the S-matrix. We note that the basic idea of a collision matrix (though not the name S-matrix which was due to Heisenberg) was introduced by John Wheeler in [108]. (See volume three of [86] for scattering theory.) Also, Cecile DeWitt-Morette spoke to us so eloquently concerning the importance of the mathematical theory of stochastic processes. (See [56] in this regard.) Finally, for the use of infinite dimensional manifolds in the geometrization of quantum mechanics we refer the reader to the recent work of T. W. B. Kibble in [60]. For much of the mathematics used here, we recommend Choquet-Bruhat, DeWitt-Morette, and Dillard-Bleick [19].

## VI. ELEMENTARY PARTICLE PHYSICS [21, 61, 64, 90]

Since the fundamental work of Eugene Wigner [112] in 1939 on unitary representations of the inhomogeneous Lorentz group = Poincaré group and continuing through the work of Murray Gell-Mann [36] and Yuval Ne'eman [77] in 1961 with SU(3) symmetry, the representation theory of Lie groups has been a cornerstone in elementary particle physics. In 1964, the quark model was proposed independently by Gell-Mann [37] and George Zweig [116]. At that time it appeared that three quarks (u = up, d = down, s = strange) sufficed to explain the spectrum of hadronic states - three quarks for a baryon and a quark-antiquark pair for a meson. Fifteen years later, it appears  these three must be supplemented by three more quarks (c = charmed, t = top, b = bottom) in order to explain recent experimental results. Thus a higher symmetry group, perhaps SU(6), is needed and one talks of quantum chromodynamics = QCD as the quarks not only come in six flavors, they come in each of three colors - red, green, and blue. (We refer the reader to F. E. Close [21] for a recent account of the quark model.) With the success of the Weinberg-Salam model for the unification of the weak and electromagnetic interactions with gauge group U(2) = SU(2) × U(1), there is hope that there is a "grand unified theory" = "GUT" which would include the strong interaction as well.  Again, some "supergroup" such as SU(6) × SU(2) × U(1) would be required.  Hopes are high, but everyone expects another "surprise" to be waiting just around the corner as energies are pushed higher and higher.  So goes particle physics.

Common to all these unification schemes is that of a symmetry group which is a certain Lie group.  The Lorentz group is a six dimensional Lie group, the Poincaré group a ten dimensional Lie group, the conformal group a fifteen dimensional Lie group, and SU(n) a Lie group of dimension $n^2 - 1$.  Thus, one speaks of Lorentz, Poincaré, and conformal invariance. It is my purpose in the next paragraph to point out a surprising (to many, but not to all!) fact about these wonderful "continuous groups" of Sophus Lie, i.e., Lie groups.

A Lie group is an algebraic - analytic - topological entity par excellence.  It consists of a smooth manifold (even a real analytic manifold) with a group structure in which the group operations are smooth (real analytic).  However, it is a remarkable fact that the differentiable structure (smooth or real analytic) follows from the algebraic and topological structure.  Let us be very precise here:  We recall that a topological space (no separation axioms assumed) is locally Euclidean iff each point in the space is contained in an open set which is homeomorphic to an open subset of $R^n$.  In 1900, at the International Congress of Mathematicians in Paris, David Hilbert proposed a list of twenty-three problems which he hoped would play an important role in the development of twentieth century mathematics.  (He could not have been more prophetic in this regard [17].)  It is the fifth of Hilbert's

problems which we have in mind.  Stated in today's language,
Hilbert's fifth problem is as follows:  Every locally Euclidean topological group is a Lie group.  Its solution required
some fifty years with an affirmative answer given by the monumental work of Gleason [39] and Montgomery-Zippin [76] in
1952.  (It was indeed fortunate that Professor Gleason was in
attendance at the Loyola conference.  His penetrating insight
proved most valuable to all of us.)  Again, it is amazing that
the differentiable structure for a Lie group follows from the
fact that we have a topological group (no separation axioms
assumed!) and an open subset of the identity element which is
homeomorphic to an open subset of $R^n$.  However, we hasten to
point out that there do exist topological (in fact, piecewise
linear) manifolds admitting no topologically compatible differentiable structure [59].  From an aesthetic point of view,
it would be most unfortunate if Lie groups did not appear as
the symmetry groups of the particle physicist.

## VII.  GAUGE FIELD THEORIES [1, 10, 27, 80]

Perhaps no place else in theoretical physics have the
techniques of modern differential geometry (in particular,
the theory of fibre bundles) been in the forefront during the
last few years as in the theory of gauge fields.  Who has not
heard of electromagnetism, the Yang-Mills theory, and the
Weinberg-Salam model mentioned as examples of gauge field theories?  In fact, to many [78], general relativity is a gauge
field theory.  In the next paragraph, we sketch the rudiments
of a "dictionary" in which we translate gauge field terminology into the language of fibre bundles.

Loosely speaking a gauge field theory is a Lagrangian

field theory which is invariant under a group of "internal

symmetries" the effect of which varies from point to point in

space-time.  Put in mathematical terms, a gauge field theory

is a connection in a (smooth) principal fibre bundle (P, M, $\pi$,

G).  The gauge group is the structure group G of the principal

bundle, the gauge potential is the connection 1-form $\omega$, and

the gauge field is the curvature 2-form $\Omega$.  The phase factor

(which, as Wu and Yang [113] have taught us, gives a complete

description of electromagnetism; however, see ([80], p. 166)

in this regard) corresponds to parallel displacement.  Parti-

cle fields are then sections of smooth vector bundles associ-

ated to the principal fibre bundle (P, M, $\pi$, G) through vari-

ous representations of the gauge group G.  A "choice of local

gauge" is just a local section s: U$\longrightarrow$P where U is some open

subset of M.

In electromagnetism and in the case of the Yang-Mills the-

ory (isotopic spin gauge field), the base space M is just

Minkowski space-time with gauge groups U(1) and SU(2), respec-

tively.  Here, the principal fibre bundles are trivial and we

have globally defined sections.  (We recall that a principal

fibre bundle admits a global section iff it is trivial, i.e.,

isomorphic to a product bundle.)  However, electromagnetism

with a magnetic monopole gives rise to a nontrivial principal

U(1)-bundle [113].  In the Weinberg-Salam unified nonabelian

gauge theory of the weak and the electromagnetic interactions

[1], the gauge group is U(2) = SU(2) $\times$ U(1).

Following Atiyah [5], we recall the mathematical framework

of the Yang-Mills equations in dimension n.  Let (P, M, $\pi$, G)

be a (smooth) principal fibre bundle with total space P, base

space M, bundle projection $\pi$: P$\longrightarrow$M, and structure group the

Lie group G.  We assume that M is oriented and carries a
Riemannian (or pseudo-Riemannian) structure.  (For instance,
M could be Minkowski space-time.)  Hence the Hodge star operator * is defined and takes p-forms into "dual" (n-p)-forms.
Now let A denote the connection 1-form on our principal bundle and let F denote its curvature 2-form.  We recall that A
and F are Lie algebra valued forms, i.e., they take their values in the Lie algebra L(G) of the Lie group G.  Thus $*$ F is
the "dual" (n-2)-form with values in L(G).  Finally, as our
Lagrangian density we choose F $\wedge$ (*F) where the exterior product is combined with the Killing form on the Lie algebra L(G)
of the Lie group G to produce an n-form on M.  The Yang-Mills
equations for our connection 1-form A are then the associated
Euler-Lagrange equations which we obtain via the usual variational principle.  For G nonabelian, the Yang-Mills equations
are second-order nonlinear partial differential equations.

As another important outgrowth resulting from the collab-
oration between physicists and mathematicians one cannot help
but marvel at the work which has been done by Atiyah, Hitchin,
and Singer [7], Atiyah and Ward [8], and Hartshorne [46], concerning classical solutions of the Yang-Mills equations using
the methods of algebraic geometry as well as complex manifolds,
Chern classes, and the Atiyah-Singer index theorem.  (We refer
the reader to Mayer in [72] for a general survey.)  Very few
physicists or mathematicians would have ever dreamed of this
spectacular synthesis of physical and mathematical ideas; yet
it is going on at this very time.  Recently Atiyah, Hitchin,
Drinfeld, and Manin [6] have found a complete set of solutions
to the self-dual Yang-Mills equations in the Euclidean metric
on the four dimensional sphere $S^4$.  This is usually referred

to as the "instanton" case.

## VIII. CONCLUSION

In the preceeding sections, we have attempted to illustrate how certain areas of mathematics (especially modern global differential geometry) have a lot to do with certain developments in theoretical physics. Of course we have omitted discussions of many topics which other writers would no doubt have included. We give a very brief discussion of two of these in the next two paragraphs.

One of these topics would almost certainly have been Roger Penrose's "Twistor Program" [83]. Penrose has voiced the opinion (as did Brans, Finkelstein, Marlow, and Wheeler at the Loyola conference and we refer the reader to their contributions to these proceedings) that the "continuum" concept should be eliminated from the foundations of physics and replaced by some structure which is purely combinatorial in nature. Thus the laws of physics should be ultimately combinatorial in nature. Here one finds a natural setting for some beautiful ideas from the theory of complex manifolds. For recent results along these lines we refer the reader to Flaherty [32] and Wells [106] and the references there.

Other important topics which we did not deal with are the recent work concerning supersymmetry, supergravity, and quantum field theory in curved space-time. Supersymmetry as formulated in 1974 by Wess and Zumino [107] is concerned with a "supersymmetry" which transforms bosons into fermions and fermions into bosons with parameters which are constant (space-time independent) spinors. Supergravity is a synonym for local supersymmetry. Here the spin two graviton which is a

boson corresponds to a spin $\frac{3}{2}$ particle called the gravitino
which is a fermion.  Supergravity has as its goal the description of general relativity in terms of the language of quantum
field theory and the ultimate unification of the four basic
interactions - the strong, the electromagnetic, the weak, and
the gravitational.  For a recent survey of these ideas, we refer the reader to Peter van Nieuwenhuizen [99] and the references there.  For a discussion of quantum field theory in
curved space-time, we refer the reader to the article by C. J.
Isham [55] which was recommended by Steve Fulling at the
Loyola conference.  (See Fulling's paper in these proceedings.)  Also, we refer the reader to the survey by Bryce
DeWitt [25] and the references there.  For recent work concerning quantum gravity and path integrals we refer the reader
to Stephen Hawking [48] and the references there.  (Also, see
the contribution by Arthur Komar in these proceedings.)

Again, let us return to the "space-time" concept.  Fundamental to both relativity theory and quantum theory is the
choice of an appropriate model for "space-time."  Roger
Penrose [83] and David Finkelstein [31] (in their twistor and
space-time code programs, respectively) opt for some type of
"combinatorial" or "discrete" structure whereas John Wheeler
[111] advocates abandoning "geometry" and substituting in its
place what he calls "pregeometry" which to him is only "an
idea for an idea" at this time.  On the other hand, Robert
Hermann [53] and Isadore Singer [9] take the view that "space-
time" should be modelled on the "continuum" and that modern
global differential geometry is of fundamental importance to
physics.  Clearly these are very different viewpoints and one

hopes that the "space-time" problem will be resolved in some satisfactory theory of quantum gravity. However, in accord with Niels Bohr's principle of complementarity [15], one might argue that "space-time" is too rich a structure to be pinned down by a single description. Perhaps several overlapping and possibly incompatible descriptions are needed to exhaust the complex variety of "space-time."

Finally, I am reminded of Albert Rothenberg's theory of "Janusian Thinking" [87]. Roughly speaking, Janusian thinking consists of bringing two or more opposite or antithetical concepts together and integrating them into some viable theory. To Rothenberg, this is what Einstein did in bringing together inertial effects and gravitational effects in the general theory of relativity. In my opinion, another example of Janusian thinking is what one is attempting to accomplish in the theory of supersymmetry. Here one is trying to bring bosons and fermions (certainly antithetical entities) under the umbrella of one unified theory. Also, this applies to the "space-time" concept where one must reconcile the combinatorial or discrete model with that of the continuum. I find Rothenberg's ideas on the "creative process" to be quite intriguing.

Let me close on a personal note. My first exposure to the point of view that fibre bundles might have something to do with physics came from reading _Fibre Bundles Associated With Space-Time_ by A. Trautman [98] and volume one of _Vector Bundles In Mathematical Physics_ by Robert Hermann [52]. These accounts were written roughly a decade ago and were, as far as I know, the earliest works advocating the use of fibre bundle techniques per se in physics. Today I stand convinced

that fibre bundles and modern global differential geometry
have a lot to do with physics and agree with C. N. Yang in
[114] where he states:  "That nonabelian gauge fields are con-
ceptually identical to ideas in the beautiful theory of fiber
bundles, developed by mathematicians without reference to the
physical world, was a great marvel to me."  Also, my first exposure to the use of differential forms and deRham's theorem
in physics came from reading John Wheeler's Geometrodynamics
[109] published in 1962.  Thus was I introduced to Wheeler's
world of physics and I have continued to follow his work to
the present time.  I cannot convey the enjoyment and excitement that his works have brought to me.  At the Loyola conference, it was my pleasure to meet Professor Wheeler for the
first time and I will always remember that event on my world
line.  In closing, let me quote from a passage by one of
Wheeler's former students, Kip Thorne, in [96].  Here, Thorne
movingly describes his first meeting with Wheeler as a graduate student at Princeton in 1962.  He captures the mood and
content of their discussion by quoting from Wheeler's 1963
Les Houches Lectures:

   "There have been few occasions in the history
   of physics when one could surmise more surely than he
   does now, in the case of gravitational collapse, that
   he confronts a new phenomenon, with a mysterious nature of its own, waiting to be unraveled ··· .  Whatever the outcome [of future investigations], one feels
   that he has at last in gravitational collapse a phenomenon where general relativity dramatically comes
   into its own, and where its fiery marriage with quantum
   physics will be consummated."

How deeply prophetic were John Wheeler's remarks in 1963 considering the monumental work by Stephen Hawking [47] on the quantum evaporation of black holes a little over a decade later. I think that Wheeler's words are just as applicable to the physics of today.

## REFERENCES


1.  E. Abers and B. W. Lee, Gauge Theories, Physics Reports 9 (1973) 1-141.

2.  R. Abraham and J. Marsden, Foundations of Mechanics, 2nd edition, Addison-Wesley, Reading, Mass., 1978.

3.  J. Anderson, Covariance, invariance, and equivalence, Gen. Rel. Grav. 2 (1971) 161-172.

4.  V. Arnold, Mathematical Methods of Classical Mechanics, Graduate Texts in Mathematics, vol. 60, Springer-Verlag, New York, N.Y., 1978.

5.  M. F. Atiyah, Geometry of Yang-Mills fields, Lecture Notes in Physics, vol. 80, pp. 216-221, Springer-Verlag, New York, N.Y., 1978.

6.  M. F. Atiyah, N. Hitchin, B. G. Drinfeld, and Yu. I. Manin, Construction of instantons, Phys. Letters 65A (1978) 185-187.

7.  M. F. Atiyah, N. J. Hitchin, and I. M. Singer, Deformations of instantons, Proc. Nat. Acad. Sci. 74 (1977) 2662-2663.

8.  M. F. Atiyah and R. S. Ward, Instantons and algebraic geometry, Commun. Math. Phys. 55 (1977) 117-124.

9.  G. Bari Kolata, Isadore Singer and differential geometry, Science 204 (1979) 933-934.

10.  J. Bernstein, Spontaneous symmetry breaking, gauge the-
     ories, Higgs mechanism and all that, Rev. Mod. Phys.
     46 (1974) 7-48.

11.  G. Birkhoff and J. von Neumann, The logic of quantum
     mechanics, Ann. of Math. 37 (1936) 823-843.

12.  A. R. Bishop and T. Schneider, editors, Solitons and
     Condensed Matter Physics, Springer Series in Solid-
     State Sciences, vol. 8, Springer-Verlag, New York,
     N.Y., 1978.

13.  K. Bleuler et al, editors, Differential Geometrical
     Methods in Mathematical Physics II, Lecture Notes in
     Mathematics, vol. 676, Springer-Verlag, New York,
     N.Y., 1978.

14.  N. N. Bogolubov, A. A. Logunov, and I. T. Todorov, In-
     troduction to Axiomatic Quantum Field Theory, W. A.
     Benjamin, Reading, Mass., 1975.

15.  N. Bohr, The quantum postulate and the recent develop-
     ment of atomic theory, Atti del Congresso Inter-
     nazionale dei Fisici, Como 1927; Nature Suppl. 121
     (1928) 78, 580.

16.  H. Bondi, Relativity theory and gravitation, Einstein:
     A Centenary Volume, edited by A. P. French, pp. 113-
     129, Harvard University Press, Cambridge, Mass.,
     1979.

17.  F. Browder, editor, Mathematical developments arising
     from Hilbert Problems, Proceedings of Symposia in
     Pure Mathematics, vol. XXVIII parts 1 and 2,
     American Mathematical Society, Providence, R. I.
     1976.

18.  L. Castell et al, editors, Quantum Theory and Structure
        of Time and Space, vol. I (1975) and vol. II (1977),
        C. Hanser Verlag, Germany.

19.  Y. Choquet-Bruhat, C. DeWitt-Morette, M. Dillard-Bleick,
        Analysis, Manifolds and Physics, North-Holland, New
        York, N.Y., 1977.

20.  C. J. S. Clarke and B. G. Schmidt, Singularities:  The
        state of the art, Gen. Rel. Grav. 8 (1977) 129-137.

21.  F. E. Close, An Introduction to Quarks and Partons,
        Academic Press, New York, N.Y., 1979.

22.  J. A. de Azcárraga, editor, Topics in Quantum Field The-
        ory and Gauge Theories, Lecture Notes in Physics,
        vol. 77, Springer-Verlag, New York, N.Y., 1978.

23.  B. d'Espagnat, editor, Foundations of Quantum Mechanics,
        Proc. of the International School of Phys. - Enrico
        Fermi, Academic Press, New York, N.Y., 1971.

24.  Dell'Antonio et al, editors, Mathematical Problems in
        Theoretical Physics, Lecture Notes in Physics,
        vol. 80, Springer-Verlag, New York, N.Y., 1978.

25.  B. DeWitt, Quantum Field Theory in Curved Spacetime,
        Physics Reports 19C (1975) 295-357.

26.  C. T. J. Dodson, Space-time edge geometry, International
        J. of Theoretical Phys. 17 (1978) 389-504.

27.  W. Drechsler and M. E. Mayer, Fiber Bundle Techniques in
        Gauge Theories, Lecture Notes in Physics, vol. 67,
        Springer-Verlag, New York, N.Y., 1977.

28.  M. J. Dupré, Geometrodynamics as foundations of physics,
        Mathematical Foundations of Quantum Mechanics, edited
        by A. R. Marlow, pp. 339-346, Academic Press, New
        York, N.Y., 1978.

29.  F. J. Dyson, Missed opportunities, Bull. Amer. Math.
         Soc. 78 (1972) 635-652.

30.  A. Einstein, L. Infeld, and B. Hoffmann, The gravita-
         tional equations and the problem of motion, Ann. of
         Math. 39 (1938) 65-100.

31.  D. Finkelstein, Space-time code IV, Phys. Rev. D 9
         (1974) 2219-2231.

32.  E. J. Flaherty, Hermitian and Kählerian Geometry in
         Relativity, Lecture Notes in Physics, vol. 46,
         Springer-Verlag, New York, N.Y., 1976.

33.  H. Flanders, Differential Forms, Academic Press, New
         York, N.Y., 1963.

34.  T. Frankel, Maxwell's equations, Amer. Math. Mon. 81
         (1974) 343-349.

35.  T. Frankel, Gravitational Curvature:  An Introduction to
         Einstein's Theory, W. H. Freeman and Company, San
         Francisco, Ca., 1979.

36.  M. Gell-Mann, The Eightfold Way:  A Theory of Strong
         Interaction Symmetry, California Institute of Tech-
         nology Laboratory Report CTSL-20 (1961), reprinted
         in The Eightfold Way by Gell-Mann and Ne'eman,
         pp. 11-57, W. A. Benjamin, Reading, Mass., 1964.

37.  M. Gell-Mann, A schematic model of baryons and mesons,
         Phys. Letters 8 (1964) 214-215.

38.  R. Geroch, Spinor structures of spacetime in general
         relativity I, J. Math. Phys. 9 (1968) 1739-1744.

39.  A. M. Gleason, Groups without small subgroups, Ann. of
         Math. 56 (1952) 193-212.

40.  A. M. Gleason, Measures on the closed subspaces of a
         Hilbert space, J. of Math. and Mech. 6 (1957) 885-893.

41.  H. Goldstein, Classical Mechanics, Addison-Wesley,
       Reading, Mass., 1959.

42.  R. J. Greechie, Orthomodular lattices admitting no
       states, J. Comb. Theory 10 (1971) 119-132.

43.  K. I. Gross, On the evolution of noncommutative harmonic
       analysis, Amer. Math. Mon. 85 (1978) 525-548.

44.  R. Haag and D. Kastler, An algebraic approach to quan-
       tum field theory, J. Math. Phys. 5 (1964) 848-861.

45.  P. R. Halmos, Introduction to Hilbert Space, 2nd edi-
       tion, Chelsea, New York, N.Y., 1957.

46.  R. Hartshorne, Stable vector bundles and instantons,
       Commun. Math. Phys. 59 (1978) 1-15.

47.  S. Hawking, Particle creation by black holes, Commun.
       Math. Phys. 43 (1975) 199-220.

48.  S. Hawking, Quantum gravity and path integrals, Phys.
       Rev. D 18 (1978) 1747-1753.

49.  S. Hawking and G. Ellis, The Large Scale Structure of
       Space-Time, Cambridge University Press, New York,
       N.Y., 1973.

50.  S. Hawking and R. Penrose, The singularities of gravita-
       tional collapse and cosmology, Proc. Roy. Soc. Lond.
       A 314 (1970) 529-548.

51.  W. Heisenberg, Nonlinear problems in physics, Physics
       Today 20 (May 1967) 27-33.

52.  R. Hermann, Vector Bundles in Mathematical Physics,
       vols. I and II, W. A. Benjamin, Reading, Mass., 1970.

53.  R. Hermann, Differential forms and Bohr-Sommerfeld quan-
       tization in general relativity, Loyola Conference on
       Quantum Theory and Gravitation, these proceedings.

54.  C. A. Hooker, editor, Contemporary Research in the
         Foundations and Philosophy of Quantum Theory, vol.
         2, D. Reidel Publishing Co., Boston, Mass., 1973.

55.  C. J. Isham, Quantum field theory in curved space-time:
         A general mathematical framework, Lecture Notes in
         Mathematics, vol. 676, pp. 459-512, Springer-Verlag,
         New York, N.Y., 1978.

56.  K. Ito, Stochastic Processes, Aarhus Lecture Notes
         Series No. 16, Aarhus Universitet Matematisk
         Institut, Aarhus, Denmark, 1969.

57.  J. Jauch, Foundations of Quantum Mechanics, Addison-
         Wesley, Reading, Mass., 1968.

58.  D. Kastler, editor, C* - Algebras and their Applications
         to Statistical Mechanics and Quantum Field Theory,
         Proc. of the International School of Phys. - Enrico
         Fermi, Academic Press, New York, N.Y., 1976.

59.  M. Kervaire, A manifold which does not admit any differ-
         entiable structure, Comment. Math. Helv. 34 (1960)
         257-270.

60.  T. W. B. Kibble, Geometrization of quantum mechanics,
         Commun. Math. Phys. 65 (1979) 189-201.

61.  J. Kokkedee, The Quark Model, W. A. Benjamin, Reading,
         Mass., 1969.

62.  A. Komar, Constraint formalism of classical mechanics,
         Phys. Rev. D 18 (1978) 1881-1886.

63.  B. Kostant, Quantization and unitary representations,
         Lecture Notes in Mathematics, vol. 170, pp. 87-208,
         Springer-Verlag, New York, N.Y., 1970.

64.  D. B. Lichtenberg, Unitary Symmetry and Elementary Par-
         ticles, 2nd edition, Academic Press, New York, N.Y.,
         1978.

65.  G. W. Mackey, Quantum mechanics and Hilbert space, Amer.
     Math. Mon. 64 (1957) 45-57.

66.  G. W. Mackey, The Mathematical Foundations of Quantum
     Mechanics, W. A. Benjamin, Reading, Mass., 1963.

67.  G. W. Mackey, Harmonic analysis as the exploitation of
     symmetry - a historical survey, History of Analysis,
     edited by R. J. Stanton and R. O. Wells, Jr.,
     pp. 73-228, Rice University Studies, Vol. 64, Nos. 2
     and 3, Spring-Summer 1978, Rice University, Houston,
     Texas.

68.  S. MacLane, Geometrical Mechanics, Parts I and II,
     Lecture Notes, Department of Mathematics, University
     of Chicago, 1968.

69.  S. MacLane, Hamiltonian mechanics and geometry, Amer.
     Math. Mon. 77 (1970) 570-586.

70.  A. R. Marlow, editor, Mathematical Foundations of Quan-
     tum Theory, Academic Press, New York, N.Y., 1978.

71.  J. E. Marsden and A. Weinstein, Book review of Geometric
     Asymptotics by Guillemin and Sternberg, and Sym-
     plectic Geometry and Fourier Analysis by Wallach,
     Bull. Amer. Math. Soc. 1 (1979) 545-553.

72.  M. E. Mayer, Characteristic classes and solutions of
     gauge theories, Lecture Notes in Mathematics, vol.
     676, pp. 81-104, Springer-Verlag, New York, N.Y.,
     1978.

73.  B. Mielnik, Geometry of quantum states, Commun. Math.
     Phys. 9 (1968) 55-80.

74.  C. Misner, K. Thorne, and J. Wheeler, Gravitation, W. H.
     Freeman and Company, San Francisco, Ca., 1973.

75.  P. Mittelstaedt, Quantum Logic, Synthese Library, vol.
         126, D. Reidel Pub. Co., Boston, Mass., 1978.

76.  D. Montgomery and L. Zippin, Small subgroups of finite-
         dimensional groups, Ann. of Math. 56 (1952) 213-241.

77.  Y. Ne'eman, Derivation of strong interactions from a
         gauge invariance, Nuclear Physics 26 (1961) 222-229.

78.  Y. Ne'eman, Gravity is the gauge theory of the parallel-
         transport:  Modification of the Poincaré group,
         Lecture Notes in Mathematics, vol. 676, pp. 189-215,
         Springer-Verlag, New York, N.Y., 1978.

79.  H. C. Ohanian, What is the principle of equivalence?,
         Am. J. Phys. 45 (1977) 903-909.

80.  L. O'Raifeartaigh, Hidden gauge symmetry, Rep. Prog.
         Phys. 42 (1979) 159-223.

81.  L. Parker, The production of elementary particles by
         strong gravitational fields, Asymptotic Structure of
         Space-Time, edited by Esposito and Witten, pp. 107-
         226, Plenum Press, New York, N.Y., 1977.

82.  R. Penrose, Techniques of Differential Topology in Rel-
         ativity, Regional Conference Series in Applied
         Mathematics, no. 7, SIAM, Philadelphia, Pa., 1972.

83.  R. Penrose and M. A. H. MacCallum, Twistor Theory:  An
         Approach to the Quantization of Fields and Space-
         Time, Physics Reports 6C (1973) 241-316.

84.  C. Piron, Axiomatique quantique, Helv. Phys. Acta 37
         (1964) 439-468.

85.  C. Piron, Foundations of Quantum Physics, W. A.
         Benjamin, Reading, Mass., 1976.

86.  M. Reed and B. Simon, Methods of Modern Mathematical
         Physics:  Vol. I - Functional Analysis (1972), Vol.

II - Fourier Analysis and Selfadjointness (1975),
Vol. III - Scattering Theory (1979), and Vol. IV -
Analysis of Operators (1978), Academic Press, New
York, N.Y.

87.  A. Rothenberg, Einstein's creative thinking and the
general theory of relativity:  A documented report,
Am. J. Psychiatry 136 (January 1979) 38-43.

88.  R. K. Sachs and H. Wu, General Relativity for Mathematicians, Graduate Texts in Mathematics, vol. 48,
Springer-Verlag, New York, N.Y., 1977.

89.  I. Segal, Postulates for general quantum mechanics, Ann.
of Math. 48 (1947) 930-948.

90.  D. J. Simms, Lie Groups and Quantum Mechanics, Lecture
Notes in Mathematics, vol. 52, Springer-Verlag, New
York, N.Y., 1968.

91.  D. J. Simms and N. M. J. Woodhouse, Lectures On Geometric Quantization, Lecture Notes in Physics, vol. 53,
Springer-Verlag, New York, N.Y., 1976.

92.  J. M. Souriau, Structure des Systèmes Dynamiques, Dunod,
Paris, 1970, (2nd edition in preparation).

93.  J. Śniatycki, Application of geometric quantization in
quantum mechanics, Lecture Notes in Mathematics,
vol. 676, pp. 357-367, Springer-Verlag, New York,
N.Y., 1978.

94.  E. Sudarshan and N. Mukunda, Classical Dynamics:  A Modern Perspective, Wiley-Interscience, New York, N.Y.,
1974.

95.  J. L. Synge, Relativity:  The General Theory, North-
Holland, Amsterdam, 1966.

96.  K. S. Thorne, Nonspherical gravitational collapse:  A
       short review, Magic Without Magic:  John Archibald
       Wheeler, edited by J. R. Klauder, p. 231, W. H.
       Freeman and Company, San Francisco, Ca., 1972.

97.  A. Trautman, Foundations and current problems of general
       relativity, Lectures On General Relativity, pp. 1-
       248, Brandeis Summer Institute in Theoretical Phys-
       ics, Prentice-Hall, Englewood Cliffs, N.J., 1965.

98.  A. Trautman, Fibre bundles associated with space-time,
       Reports on Math. Phys. 1 (1970) 29-62.

99.  P. van Nieuwenhuizen, An introduction to supergravity,
       Gen. Rel. Grav. 10 (1979) 211-226.

100.  V. S. Varadarajan, Geometry of Quantum Theory, vols. I
       and II, Van Nostrand Reinhold Co., New York, N.Y.,
       1968 and 1970.

101.  V. S. Vladimirov, Methods of the Theory of Functions of
       Many Complex Variables, M. I. T. Press, Cambridge,
       Mass., 1966.

102.  J. von Neumann, Mathematical Foundations of Quantum
       Mechanics, translated from the German by R. T.
       Beyer, Princeton University Press, Princeton, N.J.,
       1955.

103.  C. von Westenholz, Differential Forms in Mathematical
       Physics, Studies in Mathematics and its Applica-
       tions, no. 3, North-Holland, New York, N.Y., 1978.

104.  A. Walstad, The equivalence principle, Am. J. Phys. 47
       (1979) 565-566.

105.  S. Weinberg, Gravitation and Cosmology:  Principles and
       Applications of the General Theory of Relativity,
       John Wiley and Sons, New York, N.Y., 1972.

106.  R. O. Wells, Jr., Complex manifolds and mathematical
      physics, Bull. Amer. Math. Soc. 1 (1979) 296-336.

107.  J. Wess and B. Zumino, Supergauge transformations in
      four dimensions, Nuclear Physics B 70 (1974) 39-50.

108.  J. A. Wheeler, On the mathematical description of light
      nuclei by the method of resonating group structure,
      Phys. Rev. 52 (1937) 1107-1122.

109.  J. A. Wheeler, Geometrodynamics, Italian Physical Soci-
      ety, Topics of Modern Physics, vol. I, Academic
      Press, London, 1962.

110.  J. A. Wheeler, Gravitation as geometry - II, Gravitation
      and Relativity, edited by H. Y. Chiu and W. F.
      Hoffmann, W. A. Benjamin, Reading, Mass., 1964.

111.  J. A. Wheeler and C. M. Patton, Is physics legislated by
      cosmogony?, Quantum Gravity, edited by Isham,
      Penrose and Sciama, Clarendon Press, Oxford, 1975.

112.  E. P. Wigner, On unitary representations of the inhomog-
      eneous Lorentz group, Ann. of Math. 40 (1939) 149-
      204.

113.  T. T. Wu and C. N. Yang, Concept of nonintegrable phase
      factors and global formulation of gauge fields,
      Phys. Rev. D 12 (1975) 3845-3857.

114.  C. N. Yang, Magnetic monopoles, fiber bundles, and gauge
      fields, Annals of the New York Academy of Sciences
      294 (1977) 86-97.

115.  P. Yodzis, Lorentz cobordism II, Gen. Rel. Grav. 4
      (1973) 299-307.

116.  G. Zweig, CERN Preprint 8409/Th. 412 (1964), unpub-
      lished.